T3-BSE-271

The Social Constraints on Energy-Policy Implementation

The Social Constraints on Energy-Policy Implementation

Edited by
Max Neiman
Barbara J. Burt
University of California, Riverside

Lexington Books
D.C. Heath and Company
Lexington, Massachusetts
Toronto

333.79
S67

Library of Congress Cataloging in Publication Data
Main entry under title:

The social constraints on energy-policy implementation.

Includes index.
1. Energy policy—Social aspects—United States.
2. Energy policy—Economic aspects—United States.
3. Energy policy—Political aspects—United States.
4. Local government—United States.   I. Neiman, Max.
II. Burt, Barbara J.
HD9502.U52S6134   1983   333.79'0973   81-48613
ISBN 0-669-05466-6

Copyright © 1983 by D.C. Heath and Company

All rights reserved. No part of this publication may be reproduced or transmitted in any form or by any means, electronic or mechanical, including photocopy, recording, or any information storage or retrieval system, without permission in writing from the publisher.

Published simultaneously in Canada

Printed in the United States of America

International Standard Book Number: 0-669-05466-6

Library of Congress Catalog Card Number: 81-48613

*For all my parents, Bertha and Ben and Betty
and Rubien*

Max Neiman

For Michael, Stephanie, and Samuel

Barbara J. Burt

UNIVERSITY LIBRARIES
CARNEGIE-MELLON UNIVERSITY
PITTSBURGH, PENNSYLVANIA 15213

Contents

Figures and Tables

Preface

As policy analysts have often noted, the formulation and implementation of policy in the United States is incremental. That is, short of major crises and extreme compulsion to deal with a problem, most new policies are minor departures from the status quo. The reasons for this have been spelled out in numerous standard texts on public policy. This incremental form of policymaking seems to be especially prevalent in the United States; other nations, with more-unitary political systems, more-homogeneous societies, less regional variation, and smaller territories appear to be able to deal more comprehensively with their problems. The United States, on the other hand, often moves fitfully and anemically in dealing with its problems. For example, the United States lacks a national energy policy; Western Europeans have had policies since just after the Arab oil embargo in 1973.

Whether in comparing policymaking in the United States with the policy actions of other countries or in applying models of ideal policy formulation, it is often difficult to understand the variety of social factors that constrain and limit policy. Even when such constraints are realized, they are often viewed as pathologies that need to be obviated, remedied, or somehow overcome. It is common, therefore, to lament the absence of comprehensiveness in U.S. national policy. The barriers to comprehensiveness in the United States are numerous, including the usual technical considerations, lack of knowledge, lack of resources, and lack of time. But there are many others, spanning social and institutional factors.

A good example of the conflict between policy recommendation and social constraint is the call by many policymakers and researchers for national, state, and local policies to encourage greater population densities as a means toward greater efficiency of resources, including energy use, the focus of this book. Planners and scholars encourage government to deal with energy problems by promoting greater density in residential development, thereby reducing the costs of extending energy supplies, the energy costs of delivering other services (such as police and fire protection and garbage collection), and the vehicle miles traveled by residents, particularly journey-to-work trips. The U.S. population, however, has very strong antipathies to high-density living.

The economies that a planner might envision in encouraging higher population densities might pale in comparison to the citizens' perceptions of costs associated with higher density, regardless of whether such perceptions are rooted in prejudice and misinformation (for example, that higher densities mean living next to or close to minorities or that higher densities imply less privacy and more crime and deviance). Additionally, how can higher density be achieved without implying that growth that at one time was anticipated at point A will not occur or will occur there at lower densities? Clearly a policy of encouraging higher population densities will produce what the late Donald Hagman has referred to as "windfalls and wipeouts," with some investors and landowners gaining while others lose, having invested with the idea that greater densities would result. Will the losers merely accept their losses? Some of the difficulties associated with encouraging greater energy efficiency through higher residential densities are dealt with in the chapter by Johnston and Tracy, who document the impact of suburban resistance to higher population density on greater use of mass transit.

In short, this book is written with the idea that more research must be directed toward understanding the variety of important constraints on energy policymaking. Enormous amounts of material have been generated in the past decade about the shape of optimal policy, according to a variety of perspectives. It is our view, however, that more research should be directed not only to formulating yet some other version of optimal energy policy but also to examining the important features of society that impose constraints on choice and limit the range of policy options available, not only for managing energy issues, but for managing other policy areas as well.

In chapter 1, Michael Reagan addresses an analysis of the contributions and burdens that federalism makes on the formulation of energy policy. While federalism is shown to complicate energy-policy formulation, Reagan warns against overemphasizing the importance of federalism. He points out that society's lack of consensus regarding the importance and duration of energy problems, as well as alternative notions of equity, complicate matters as much as the governing institutions.

The scale of social organization has often been seen as a major problem in dealing innovatively, creatively, and quickly with problems. Much has been made of the concentration of economic power in the private sector. Concerning energy matters, the argument is often made that large energy-related firms have prevented the introduction of new technologies, which presumably would have provided a quicker and more effective response to energy problems. On the one hand, some claim that large energy firms tend to restrain product innovation. Thomas Dietz and James Hawley in chapter 2 analyze the effect of economic concentration on the diffusion of innovative technology, focusing on the case of photovoltaic cells. The theme of bigness is also addressed in chapter 6 by Michael Sheehan, who examines the lack of political responsiveness, as well as the absence of adequate innovation among large energy suppliers. The analysis of the relationship between size and adequate performance is a preoccupation of those who believe that one major component of energy policy should be the break up of concentrations of economic, and hence political, power.

Those involved in reforming energy policy often believe that defenders of conventional or dominant energy modes do not appreciate the economic advantages of alternative, so-called unconventional, energy sources. It is probably true that scant attention has been paid to the real costs of conventional energy service. For example, it is commonplace to claim that much of our expenditures for the navy are related to the need to ensure oil supplies from certain parts of the world. Should a portion of the defense budget then be considered an energy cost or purely a military expenditure? Two of the chapters in this book are directed to reformers of energy policy who might be in danger of ignoring the limits or costs of energy reform. There is a growing tendency for advocates of mass transit, solar energy, and the breaking up of energy firms to ignore the cost of their proposals and thus not to appreciate the limits of their designs. In chapter 3, Walter Mead and Gregory Pickett evaluate two of the alternative bidding systems for leasing publicly owned oil and gas resources: (1) cash-bonus bidding with a fixed royalty and (2) profit-share bidding. Congress has identified several conflicting objectives for its leasing program. It would therefore be impossible to provide a consistent economic evaluation of alternative bidding systems using the various leasing objectives enumerated by Congress. In the present study, the authors examine how well each bidding system serves to maximize government collection of the present value of the pure economic rent derived from oil and gas production.

From an economic-theory perspective, the authors find a pure cash-bonus bidding system to be superior to profit-share bidding. The overriding advantage of bonus bidding is that it harmonizes private incentives to minimize costs relative to any revenue stream with the social welfare goal of maximizing the economic rent. From an empirical perspective the authors conclude that oil and gas leases issued by the government between 1954 and 1969 have been effectively competitive and that the government collected at least 100 percent of the available economic rent.

Similarly Karl Hausker and Eugene Bardach, in chapter 4, analyze the movement to encourage greater use of solar energy, particularly in the Southwest and especially in California. The authors implicitly criticize those who zealously press for the use of solar energy without regard to whether this soft path is really more beneficial than conventional energy delivery. Essentially Hausker and Bardach specify a set of conditions that indicate when solar water heating is actually cost effective for consumers.

The role of local government in managing the energy problem is another theme that has been studied. Because of the putative closeness of local officials to citizens, local governments increasingly are regarded as the proper governmental unit for managing many energy difficulties. When the long gasoline lines occur, citizens complain to the city hall and county offices, not to the Department of Energy in Washington. Moreover, the diversity of local conditions in climate, the energy mix used, and economic base make it difficult to develop useful policy at more-distant levels. Several of the chapters deal more directly with the concern for local government's role in energy management. In chapter 5, Robert Johnston and Steve Tracy point to the impact of decentralization of land-use policy on efforts to achieve energy efficiency through greater use of mass transit. They are somewhat pessimistic about the system of local reliance in this connection. On the other hand, Michael Sheehan, in chapter 6, is quite optimistic about the capacity of small-scale producers of energy, including locally owned utilities, to innovate and respond effectively to the nation's energy needs. Sheehan claims that a major impediment to effective and equitable energy policy is the existence of large-scale, uncompetitive producers; this chapter overlaps in substance, although not in conclusions, with the chapter by Dietz and Hawley. Sheehan's major concern is to build a case for public policies designed to assist small-scale energy producers in competing in the energy production and distribution system. He sees the assistance not as government largesse but as necessary to overcome years of policies providing uneconomic, unfair, and ineffective assistance to large actors in the energy field.

In chapter 7, Beverly Cigler summarizes the case for greater local-government involvement in energy policy. She provides some findings clarifying the conditions that facilitate such local-government action. In chapter 8, we present some findings suggesting the presence of attitudinal bulwarks against local government's taking a more active role to promote local energy conservation, particularly if that role involves a substantial amount of police-power regulation. Conservation, which is often claimed as the paramount objective of local energy policy, can be a minefield for officials since it is likely that conservation, if taken seriously, may generate a variety of issues concerning the proper role of government in interfering with private choices.

The final chapter, by Gerry Riposa, focuses on energy-policy change as an innovation process. Riposa contends that one can understand the prospects for and limits on policy change by looking at the innovation process generically and then applying some of the more salient findings to produce hypotheses that can account for energy-policy innovation.

Acknowledgments

For assistance in completing various stages of this project, we extend our gratitude to the University-wide Energy Research Group of the University of California, the Academic Senate Committee on Research, the Center for Social and Behavioral Sciences Research, and the Energy Sciences Program at the University of California, Riverside.

1 The Federalism Factor in Energy Policy

Michael D. Reagan

Although the Reagan administration seems not to want an energy policy, as indicated by its efforts to dismantle the Department of Energy and its announced plan for almost total reliance on the private marketplace to take care of national energy needs (NEPP 1981; Reagan 1982), an energy policy—fragmented and incomplete—does exist as the result of presidential and congressional actions dating back to the immediate post-World War II years and culminating in the intense efforts and partial accomplishments of the Carter administration (Goodwin 1981). Presidential-congressional-bureaucratic-interest groups have created an enormously complex web of statutes and implementing programs at the national level. Greater yet is the complexity of energy policy if we look into the added dimensions imposed by federalism. The basic purpose of this chapter is to examine federalism as both a constraint upon and an encouragement toward effective U.S. energy policies. The orienting premise is that governmental structure and policy issue affect one another; sometimes one is the dependent variable, sometimes the other.

The analysis is organized around four ways in which the federal factor interacts with energy policy:

1. Impacts flowing from the structural fact that ours is a federal system whether defined as a constitutional given or as simply a political fact of permissive federalism (Reagan and Sanzone 1981).
2. Substantive content of national laws that either restrict state options or mandate specific policies onto the states.
3. The opportunity created by federalism for the states to act as innovative laboratories, devising programs that may later be incorporated into national policy and thus diffused among all the states, and for the national government to assist the states in planning.
4. Impacts of state policies and state independence upon national energy-policy development, especially ones that restrict the national government's options, an element of the reciprocal nature of federalism that is not always sufficiently attended to in examinations of national policy development.

The Inevitabilities of Federal Structure

No matter how valiant the efforts of the current national administration to turn functions back to the states and to diminish the federal role, it is safe to say that the range of areas in which policy and program authority are shared between the national and state levels will remain vastly greater than that in which the national role is eliminated. The shared-functions approach to public policy may not be an inherent result of federalism; the "marble cake" (Grodzins 1966) may once really have been a "layer cake." The fact of positive government at the national level is now irreversible overall, no matter how the exact outline may vary from time to time. Three basic factors support this summary judgment. First, modern communications and transportation create a nationwide and largely uniform sense of national problems, the means by which they may be attacked, and the means by which effective political coalitions may be formed to place them on the federal agenda. Second, economic and technological interdependence render state autarky technically impossible. Third, public attitudes toward the role of government (the expectations that Washington will act) reflect permanent recognition of the first two facts of life. In this regard, the Employment Act of 1946 may be cited as the symbolic benchmark by declaring the national government's responsibility to promote full employment.

Within this context, it is not surprising to find that a topic as broad as energy is now inevitably an area of shared responsibility and authority. (For a succinct and useful review of the steps by which the present situation got to where it is, see Aron 1979.) Let us examine a few of the resultant areas of reciprocal impact.

Land-Use Planning

The national government (and Indian tribes) own almost 45 percent of the land in eight western states (Arizona, Colorado, Montana, New Mexico, North Dakota, South Dakota, Utah, and Wyoming). Within this area lie 37 percent of the U.S. total of coal reserves, 90 percent of the uranium, and virtually all of the oil shale, according to a 1977 estimate. Putting these two strands together the national government owns about half the coal and uranium and 80 percent of the oil shale in that area (Hall, White, and Ballard 1978) it can hardly escape being a dominant determiner of resource development in those states. A related factor that affects internally the national government's choices in this area is the impact that energy-resource development might have on the aesthetic enjoyment of some prominent national parks, including Yellowstone, Bryce, Zion, and Grand Canyon, and on the unusually high level of environmental quality enjoyed generally in those states.

Resource Management

Whether national policy stresses development of domestic energy resources to maximize immediate independence or importation to maximize conservation of our own reserves for use at a later time is a decision that directly affects the rate of development within the states. Whether the national government closely supervises the extractive process or leaves that function to the several states is also a major question.

Water is the quintessential resource in the western states. Large quantities of it are needed for some energy-development processes, such as shale-oil extraction. Federal-state interdependence in this policy area is long established, ranging from the amount of federal water allowed per farm to the approval of interstate compacts for water development and distribution patterns and the licensing of power dams.

Research and Development

Since World War II (and with scattered earlier instances), the national government has played a substantial role in funding and directing research and development (R&D), not least in relationship to energy resources, and most notably within that its unique role in atomic energy, began as a total government monopoly (Dupree 1957; Reagan 1969; Goodwin 1981). One of the most far-reaching choices facing U.S. energy policymakers is whether to provide developmental subsidies (whether as grants, recourse-free loans, or guaranteed purchases) for synthetic fuels. An on again-off again record exists up to now regarding national funding and private enterprise's interest in utilizing that funding, the trials and tribulations of the Exxon-Tosco shale-oil project being a prime example (*Wall Street Journal,* May 5, 1982; *Los Angeles Times,* May 3, 1982). It now seems clear that without promotional subsidy, the pace will be very slow; with it, it might be very rapid. Rapid development will have substantial consequences for the states involved, in matters ranging from tax revenues to boomtown infrastructure expenses. There is little evidence, however, that the states will have much of a voice in the basic policy decisions. Certainly they did not in either the crash development of the Carter administration's plans for the Synthetic Fuels Corporation or the Reagan administration's substantial pullback from that overblown initiative.

Environmental Protection

Environmental protection constitutes another realm of shared authority with substantial energy-policy implications. Indeed energy policy and environmental policy are often reverse sides of a coin, sometimes working together but often contesting with one another for policy priority. While we are perhaps most conscious of Washington's role as a generally stronger upholder of environmental values than state governments under more-immediate business-development pressures, the near-panic reaction to the second oil crisis of 1979 produced legislation (eventually aborted) that would have created an Energy Mobilization Board with power to supersede environmentally protective state policies in the name of providing a fast track for approved energy development projects (CQ, 1981, pp. 230–233). The federal relationship is sometimes one of threats to preempt established state policies that had earlier been ardently encouraged by national administrations and national legislation. Air- and water-quality programs and strip-mining legislation come to mind as examples of some of the dimensions in which power is shared in an explicitly federalist manner.

Interstate and International Considerations

Acid rain may be created by midwestern power plants and fall on eastern states or eastern Canada. The power production is an intrastate matter; the problem created is one that involves the national jurisdiction. Montana's 30 percent severance tax on coal creates interstate tension with consumers in neighboring states whose utility bills reflect their utilities' added fuel costs.

Energy-Contingency Planning

Odd-even gasoline lines were a matter for gubernatorial prerogative (or a gubernatorial burden) during an oil disruption; national authority for contingency programs was allowed to lapse (under urging from the administration) in September 1981, although some emergency plans do continue in place, such as the Strategic Petroleum Reserve. Situations being clearly quite different regarding oil and gas supply and use patterns in the various regions of the United States, it is logical that substantial state option be permitted in responding to any energy disruption. On the other hand, because oil-importation dependence is a national fact, Washington has some basic responsibility for contingency planning. Melding the two is currently a matter of dispute between the administration and the Congress.

Nuclear Power

Leaving aside the international dimensions ranging from overseas sales of reactors to nuclear proliferation as a factor in the arms pattern of the world, domestic nuclear power is one of the most hotly debated, emotionally volatile of issues in the energy-policy arena. Because citizen groups and state utility commissions have become embroiled in the pros and cons of nuclear power and because investor-owned utilities are under the jurisdiction of state utility commissions while the Nuclear Regulatory Commission has basic responsibility in the law and in the eyes of the public for nuclear licensing and power-plant safety, this is an area of extremely sensitive and complicated intergovernmental relations. Nuclear power is the only energy source whose utilization began as a totally national, socialized (government-owned) enterprise, with the 1954 amendments to the atomic energy legislation of 1946 only partially devolving nuclear-power development onto the private sector. State regulation came later, whereas it has typically preceded national action in most other spheres of domestic policy. A current question of federalism is the extent to which national atomic energy legislation does or does not preempt state regulation.

The areas discussed do not provide an exhaustive cataloging of federalism-affected areas of energy policy, but it should be sufficient to underscore the proposition that taking the world of energy as it now is in terms of technology, economics, politics, and public opinion, almost any major issue of energy policy, whether arising at the state or the national level, will involve questions of federalist import: questions of appropriate loci of authority and questions of substantive capability for resolving a given problem.

National Policies as Constraints on State Policymaking

Specific national policies—mostly statutory measures—constrain state policymakers by limiting their discretion, mandate specific policies upon the states, or use the states as implementers of nationally funded energy programs. The coverage presented here is illustrative of the high spots rather than a complete listing of the contents of each category. Some policies are energy policies in the direct sense of having been designed primarily from an energy perspective, while others were designed with other considerations primary (mostly environmental protection) but have substantial impacts on energy resources and energy policy.

Natural Gas Pricing: Control and Decontrol

The Natural Gas Policy Act of 1978 (PL 95-621; Stat. 3550) established a phased pattern for deregulation of price controls on interstate sales, with complete decontrol to occur in 1985. It also, however, increased price control by bringing intrastate gas sales under regulation for the first time (to solve the problem of producer reluctance to sell out of state when the uncontrolled price was higher in state). Although these steps are matters of national price control rather than of federal relationships on their face, the indirect impact upon state regulatory bodies is considerable. These utility commissions review rate-increase requests made by gas and electric companies. Gas consumers are now discovering that state utility commissions are virtually powerless to protect them from large, sudden increases because the federal decontrol program allows the producers to charge more to pipeline companies, and the latter pass the increases on to the utilities in the states they serve. Fifty-two percent of all natural gas used in California in 1981, for example, came from El Paso Natural Gas of Houston, Texas. While most of that was still price controlled, 6 percent of the volume was accounted for by so-called deep gas, which is free of regulation and priced high enough to account for 17 percent of El Paso's purchase expenditures. When Southern California Gas Company customers protested, the public utility commission could only join in the protests, seconding the gas company's pleas to the pipeline company to bargain harder with the producers (*Los Angeles Times,* April 13, 1982). State policy-making with regard to gas rates is thus severely constrained by the national policy of price decontrol—necessary though that may be in the total energy picture.

National Determination of Utility Fuel Patterns

In the name of national energy self-sufficiency, the Congress passed in 1978 the Power Plant and Industrial Fuel Usage Act (PIFUA) (PL 95-620; 92 Stat. 3289). The primary provisions of this legislation promote the substitution of coal for oil or gas in power plants and large industrial boilers. The act forbids the use of natural gas or oil as the primary fuel in any new power plant or in existing plants after 1990, with some qualifications. Although the basic thrust substantially intrudes upon an area normally left to state utility commission determination, there are exemption provisions in whose operation the states are given a significant role. The act permits a permanent exemption if a coal or synthetic facility is not feasible because of state and local requirements based on the public interest, and no exemption will be issued without the approval of the state regulatory authority.

Another section constrains state regulators in promoting industrial cogeneration because of its allegedly overly complex requirements. A California Energy Commission report comments on this and provides an example of intergovernmental efforts to solve a dual-regulation problem:

> PIFUA ... provides specific exemptions for cogeneration. The most useful of these exemptions is available only when the potential cogenerator demonstrates that economic and other benefits of cogeneration are unobtainable unless oil and gas are used. To make this required demonstration, the cogen-

erator now must essentially present a very detailed, sophisticated, and difficult analysis of long-term future utility oil and gas use which might be reduced
through cogeneration. The state's cogeneration task force, a state interagency
group, is working with the federal agency that administers the cogeneration
exemptions to simplify the exemption requirements and encourage cogeneration. If these efforts fail, Congress will need to act to remove cogeneration
entirely from the prohibitions of PIFUA. [CEC 1981a, p. 107]

As with gas-price decontrol, the primary state impact of PIFUA consists
of limiting the discretion of state regulatory bodies to work out with energy
producers a pattern of fuel usage they may deem best suited to their own situation—"situation" having political as well as technological elements. However, the exemption provisions are sufficient, it seems, to provide room for
much intergovernmental bargaining in the implementation of the act. One
guesses that its ultimate extent of impact may depend more on the environmental mood of the nation toward burning coal and on the twists and turns
of the world petroleum picture than on the simple fact of having a statute.

An interesting footnote to this legislation concerns an apparent sweet-
ener to coal-producing states concerned about the public-sector costs of the
boomtown syndrome that might accompany a forced increase in demand for
coal. The act provides "impact assistance" funding to states whose governors
can show a probable 80 percent increase in coal-production employment,
out-stripping state ability to handle the infrastructure needs generated by the
population bulge. This is a type of intergovernmental compensation for costs
created in selected states by virtue of national energy policies that might be
broadened to prevent a few energy-resource-producing states from bearing a
disproportionate share of the national energy burden. A 1978 Department of
Energy study spells out a range of such impact-assistance measures that might
be considered as part of a national-state partnership (DOE 1978; see also
Plummer 1977).

Nuclear-Power-Plant Siting and Licensing

An area in which the regulatory dimension of energy policy is prominent is
that of nuclear-power-plant development. The question of whether the proposed additional power is needed and financially justified is handled by state
regulatory bodies, usually called public utility commissions. They also deal
with questions of siting. Regulation of radioactive hazards began as an area of
exclusive national jurisdiction and remains close to that, despite 1977 amendments to the Clean Air Act that permit states to regulate air emissions of radioactive materials (Aron 1979, p. 464), and consideration of environmental impacts is very much a matter of shared—often duplicated—powers. In the
heyday of the Atomic Energy Commission-Joint Committee on Atomic Ener-

gy subgovernment, environmental impact was not much of an obstacle to Atomic Energy Commission (AEC) license approval, until the *Calvert Cliffs* decision in 1971 (449 F. 2^d 1109; 1111 D.C. Cir. 1971), in which the Second U.S. Circuit Court of Appeals insisted that AEC was bound, under the National Environmental Policy Act, to do thorough environmental impact statements (EISs) on a wide variety of factors. Between that decision and early 1980, the AEC and its successor agency, the Nuclear Regulatory Commission (NRC), performed 140 EISs and cooperated in varying degrees and ever-changing ways with an increasing number of state regulatory bodies also doing EISs on the same proposed plants. The 1970-1980 decade was the period of burgeoning environmental-protection activity in most of the states.

From initial proposal by a utility through the NRC's final licensing decision, the development of a nuclear-power plant is very much a federally shared authority situation. Because of the overwhelming weight given to radioactive safety questions (even before Three Mile Island), the final authority for licensing rests with the NRC and thus constitutes a significant restraint on state discretion, although a strong case can be made that the major costs and benefits are state specific and that the value structure of the politically effective citizenry of a given state should be given great weight. At the same time, there are substantial national interests involved: health and safety, national security, energy conservation, resources planning. The balancing act continues to be difficult and all decisions controversial, as the names San Onofre and Diablo Canyon will attest in the minds of the current attentive public regarding nuclear matters. (For a thoughtful review of the national and state considerations involved and much information regarding the great variety of state licensing programs and federal-state efforts to achieve less-duplicative, more-efficient licensing reviews, see Spangler 1980.)

One potentially significant recent development if upheld by the U.S. Supreme Court, from a federalism perspective, is an October 1981 decision, *Pacific Legal Foundation v. State Energy Resources Conservation and Development Commission,* 659 F. 2d 903 (1981), by the Ninth U.S. Circuit Court of Appeals holding California's right to claim state jurisdiction over nuclear-power-plant siting with respect to certain matters other than radiation-hazard protection. In 1976 the California legislature (to ward off an even more stringent initiative measure) decreed that no new nuclear plants be built until there was a federally certified waste-disposal system (which does not even now exist) and that three sites must be presented for state regulatory review. The court sustained these statutes on the reported basis that they were made on land-use and economic criteria rather than as protection from radiation hazards.

Another volatile issue of potential federal-state conflict is currently pending: the question of where storage of radioactive nuclear wastes will take place once the national government decides upon a method and a program for permanent waste storage. As of early 1979, some 25 million gallons of radioactive sludge were stored at the Savannah River weapons plant in South Carolina, some of it shipped in from other states. The state's governor was reported to be increasingly edgy about South Carolina's distinction as being in the forefront of radwaste storage. In fact several states have recently been raising objections against even the shipment of radwaste across their land. Can the national government force a storage location on an unwilling state? Legally, probably yes; politically, probably no. Can a stalemate be avoided? Possibly, with great skill on the part of negotiators and compelling evidence of adequate safety precautions. (On the entire range of nuclear waste-management policy questions and their intergovernmental complexities, see Colglazier 1982; Kearney and Garey 1982; and Shapiro 1981.)

Energy versus Coastal Protection

Another area of strong shared jurisdiction (and strong emotions) is that of oil and gas exploration and production in the offshore area known as the outer continental shelf (OCS). The national government holds title to the OCS, and the secretary of the interior is the party authorized to sell leases for private exploration and production. Two national laws require that the secretary conduct such lease sales in close conjunction with the affected states: Coastal Zone Management Act (CZMA) of 1972 (PL 92-583; 86 Stat. 1280) and its 1976 amendments (PL 94-370; 90 Stat. 1013), and the Outer Continental Shelf Lands Act of 1953 (PL 83-212; 67 Stat. 462) and its 1978 amendments (PL 95-372; 92 Stat. 629).

CZMA requires that federal leases be certified by the affected state as consistent with its coastal zone program, although it also holds that the states are to operate their coastal zone programs "in cooperation with state and local governments." The other set of OCS statutes directs the secretary of the Interior to solicit recommendations from affected governors before leasing and to provide reasons publicly in case of rejection of the gubernatorial advice. On August 12, 1982, the Ninth U.S. Circuit Court of Appeals held that Secretary of the Interior James Watt could not carry through on a lease sale auction held in May 1981 until he had made a determination that the sales would be "consistent, to the maximum extent practicable, with the California coastal zone management plan." The court added, however, that the state did not have an absolute veto over national oil and gas actions in the OCS area. The secretary had contended that no consistency determination was required because the particular tracts would not directly affect the California coastal zone. Both the district court (in July 1981) and the circuit court rejected that contention (*Los Angeles Times,* August 13, 1982). A set of mutual constraints appears to exist in this area of federal sharing: the states affected are given explicit rights to play a role in determining the conditions under which oil and gas development will take place in coastal zones they manage by virtue of federal law, while the national government can act with finality on leases but must seek advice and must respond to it in a reasoned way if it disagrees.

Oil-Disruption Contingency Planning

In this area, the Congress and Presidents Carter and Reagan have together made an impossible set of demands upon the state governments and have, through a national-policy vacuum, constrained the states from effectively planning for themselves. I refer specifically to the requirements imposed on the states to fulfill the nation's obligations to the International Energy Agency's (IEA) Emergency Sharii. System (ESS). Provisions for this system commit member governments (inciuding the U.S. and most Western European governments, plus Canada, Japan, Australia, and New Zealand) to have standby plans for reducing their final oil-consumption levels if an emergency situation is declared by IEA. Two trigger points are set at 7 and 12 percent, respectively. When the IEA governing system determines a shortfall at either of those levels and activates the agreement, the member nations are to put into effect the consumption-reducing plans they have prepared, also at those two levels.

In the United States, the 1974 IEA agreement was followed by demand-resistant measures undertaken by the Ford administration but shelved by Carter. In early 1979, DOE proposed a new set of contingency plans to Congress, which were rejected in favor of new legislation. This took the form of the Emergency Energy Conservation Act of 1979 (EECA) (PL 96-102; 93 Stat. 749). Its terms direct each state to develop its own energy-conservation plan for emergencies and authorized the president to set monthly oil-reduction targets when declaring an actual or potential severe disruption—either independently of IEA or in accord with the ESS 7 or 12 percent trigger point. Upon such actions by the president, each state is to submit its plan within forty-five days; within thirty days more, the secretary of energy is to approve or disapprove. If after ninety days, any state is not meeting its target, a federal conservation plan may be imposed on that state (CQ 1981, p. 218; CEC 1981c, pp. K-50–51). Since the expiration in September 1981 of earlier contingency-planning statutes, this is the only nonmarket policy extant for handling an oil-supply disruption. This policy places the entire initial onus on the states and discourages them from even attempting to fulfill the obligation by telling them (since January 1981) that the market will resolve any supply-shortage problems.

In August 1982 (PL 97-229, as reported in the *National Journal*, August 7, 1982) Congress, as part of extending the antitrust waiver that enables U.S. oil companies to cooperate in the IEA emergency-sharing system, required the administration to submit to Congress its options for using the Strategic Petroleum Reserve (SPR) in time of emergency, to put 220,000 barrels a day into the SPR, and to submit to Congress a plan for responding to a severe energy-supply disruption.

Not only is the United States clearly without any firm way to meet its IEA obligations, its failure to develop a national contingency plan leaves the states entirely without a framework within which they could effectively do their own planning. The following paragraph from a California Energy Commission report illustrates the difficulties:

> California currently (1981) receives approximately 44 percent of its crude oil from in-state production, 45 percent from Alaska, and about 11 percent from Indonesia. Furthermore, oil pipelines serving states east of the Rocky Mountains are not connected to the western states—that is, Petroleum Administration for Defense District V (PADD V)—oil distribution system. But we are not totally isolated. During a shortage of oil from the Middle East, oil companies would most likely attempt to equalize supplies nationwide by diverting Alaskan and possibly Indonesian crude to Gulf Coast refineries. And if they did not, the government probably would do so rather than allowing a surplus to accrue on the West Coast while shortages affected the East and the Midwest. Similarly, if the International Energy Agency's (IEA) emergency crude sharing system is triggered by a 7 percent or greater shortfall in any member country, then portions of Alaskan and Indonesian crude oil supplies could conceivably be diverted to Japan, while domestic crude redistribution and military fuel requirements take additional portions of California's supplies. Moreover, lack of national preparation of an oil supply disruption . . . suggests that this crude redistribution process at the domestic and international levels, as well as the domestic economic effects of a shortage, may be handled haphazardly by the federal government. This situation represents yet another aspect of California's vulnerability. [CEC 1981b, pp. 2–3]

Federally Mandated Programs

The entire federal grant-in-aid system (apart from general revenue sharing) can be described as five hundred mandates; each grant program provides funds from the national level for doing something at the state or local level. Many are "carrot" programs: they provide incentives for doing something while not formally requiring that it be done; but sometimes the "carrot" is close to a "stick," as when highway funds are cut if the state does not adopt a beautification program. Energy has not escaped the mandate mode by which American politics pursues goals of the national government through the instrumentalities of the states and local governments. A few examples will illustrate.

Under the 1978 National Energy Conservation Policy Act (PL 95-619; 92 Stat. 3206), DOE created the Residential Conservation Service (RCS) program. It mandates each state to develop a detailed plan for encouraging installation of energy-conserving equipment and for taking renewable-resource measures in existing residential structures. Under DOE guidelines, the states can tailor programs to their specific needs in terms of climate, types of equipment most useful, and so forth, although state plans must be submitted to DOE for approval before federal funds are awarded. It is under this program that home-energy audits have become a commonplace activity offered by electric and gas utilities in many jurisdictions, along with low-cost conservation loans by the utilities to their customers. The tiers of implementation includes the DOE guidelines, state public utility commission plans; and utility-to-customer actions, as mandated in the respective state plans.

A second major mandate program is one that implements a congressional objective of reducing electric utilities' needs to build additional central power plants requiring huge capital outlays. This is pursued by requiring that state utility commissions in turn mandate upon the power companies a program for compulsory purchase by them of independently produced small quantities of power. The sources of such power might be cogeneration or such renewables as small hydro, geothermal, wind, solar, or biomass. In some states, an entrepreneurial boom has developed in response. Under this legislation (Public Utilities Regulatory Policies Act of 1978; PL 95-617; 92 Stat. 3117), state utility commissions are also mandated to consider with appropriate public hearings, such conservation-conducive steps in utility rate setting as prohibiting use of declining block rates and encouraging use of time-of-day rates that would reflect the higher costs of peak load power, and other special load-management techniques. The act sweetened this mandate by authorizing federal grants that the states could use in implementing the special rate studies entailed in implementing the program.

Does PURPA call for constitutionally excessive intrusion into the authority of the states by compelling them to enforce the purchase of small power by utilities and by specifying things they are to consider? The Supreme Court has answered negatively (*Federal Energy Regulatory Commission v. Mississippi,* 72 L. Ed. 2d 532, decided June 1, 1982), stating that the act is an ordinary use of the commerce clause that simply establishes "requirements for continued state activity in an otherwise preemptible field." Justice Sandra O'Connor's dissent (joined by Chief Justice Warren Burger and Justice William Rehnquist) would have rejected the challenged portions of the act on the basis that "state legislative and administrative bodies are not field offices of the national bureaucracy. Nor are they think tanks to which Congress may assign problems for extended study." By a five to four vote (Justice Lewis Powell dissented separately), PURPA has been upheld.

A slightly different kind of mandate (perhaps even stretching the usual meaning of the term) is found in the Surface Mining Control and Reclamation Act of 1977 (PL 95-87; 91 Stat. 445). This act establishes national environmental protection standards for strip-mining but permits the states to assume exclusive jurisdiction for the regulations if they wish to do so, provided they submit to the secretary of the interior a state program demonstrating capability of carrying out the nationally specified provisions. These include a state law embodying the national standards, provisions for sanctions, and "a State regulatory authority with sufficient administrative and technical personnel and sufficient funding to enable the State to regulate" adequately. A federal program is to be promulgated for each state that has not submitted an approvable program within a certain time interval. As of the end of July 1982, the twenty-four largest coal-producing states had approved plans and therefore exclusive jurisdiction over strip-mining and reclamation within their borders (*Los Angeles Times,* August 4, 1982). This seems to be a coming development in intergovernmental relations (Dubnick 1982; Dubnick and Gitelson 1981).

State Initiatives and National Cooperation

In addition to constraining the states, our federal system enables the states to act as innovative laboratories in devising energy programs of their own, and sometimes the national government assists the states. Among the state innovations are solar tax credits, R&D projects for renewables, and energy resource-development plans. There has been extensive activity in the field of nonmandated measures (Regens 1980). For state experimentation to have maximum payoff, however, it is often necessary that the national level adopt and write into its mandates the more-successful state innovations, and that is exactly what DOE has done in such areas as conservation: "We've stolen a lot of our legislation from the states, especially California," said a DOE administrator in 1980 (*Los Angeles Times,* May 18, 1980). In addition to stealing ideas, what can the national government do to assist the states in this area? Wilbanks (1981) suggests that the federal role should be to "provide more incentives for community energy planning, make information and technical help available, and perhaps assist in local institution-building."

National assistance can also take legislative form. An example is the Pacific Northwest Electrical Power Planning and Conservation Act of 1980 (PL 96-501; 94 Stat. 2697), under which Washington, Oregon, Montana, and Idaho engage in regional energy planning and development, cooperating with the national government's Bonneville Power Administration (Lee 1981). Another, currently under congressional consideration, is proposed legislation to permit use of a national right of eminent domain to make possible development of coal-slurry pipelines in the western states (utilizing, in part, railroad rights of way; railroad concerns about the competition are the reason why the legislation has been in pending status for a long time)(*National Journal,* May 8, 1982, and later updates).

These brief examples should establish the point that federalism is not always a barrier; it can also be an advantage in policy development because it provides multiple access points for ideas to take hold.

State Constraints on National Energy Developments

In recognition of the realities of the American system, we need to take note of the power of the states (and even of municipalities) to restrict the national government's options—not through constitutional impediments but through political digging in of heels. One example is that of nuclear-waste disposal: the states can at least make it difficult for Washington policymakers by simply refusing to accept federal designation of a disposal site. Another example is the negative reaction of cities to the Federal Emergency Management Administration's efforts to establish local crisis-relocation plans in the event of nuclear war. In California, Massachusetts, and Vermont particularly, a number of counties and cities have refused to participate in the program (*Los Angeles Times,* August 9, 1982).

Conclusions

What does this descriptive recital of national-state interactions in energy-policy development add up to? Our speculative response has to be in two parts: the significance of the matters covered for energy development and the significance for federalism as practiced in the United States.

Energy-policy formulation suffers most from lack of societal consensus about the importance and duration of the problem and about the equity implications of various approaches rather than from structural features of government. It is clear, however, that federalism does complicate policymaking in this field beyond what one would expect if ours were a unitary form of government. Having the national government bribe, weedle, cajole, and threaten the states is an inefficient, inelegant, and occasionally ineffective way to handle what is clearly a national problem in terms of its origins and the imperatives of amelioration. In a more strongly national system, those measures deemed necessary by the Congress would simply be administered by field offices of the DOE. Implementation would be national and probably more consistent than when the applied policymaking dimension of what has been legislated has to be shared with fifty jurisdictions subject to fifty other sets of local pressures. In such respects, energy policy (at least in the absence of a perceived urgent crisis) exemplifies the standard features of fragmentation attributable to federalism throughout the sphere of domestic policymaking.

Does energy as a policy issue have a noticeable impact upon federalism as structure and process? I think the answer clearly is yes; and the direction of impact is toward centralization. That is, the imperatives of the problem call for national action, and as national actions are taken, the balance of policy-making authority shifts further toward the national level. Whether the question is synfuel R&D subsidies, stimulating all electric utilities to maximize utilization of renewable energy sources, or legislating the conditions of trade-off between increased coal development to cut oil use and increased strip-mining damage to the physical and aesthetic environment of the western states, only the national government is in a position to make effective policy. What the nation does cannot simply be the sum of varied state policies—or their absence.

That the balance is a touchy one is illustrated by the five to four Supreme Court vote in the PURPA case. The minority is saying that energy should not push federalism so far from its earlier traditions. But in energy as elsewhere, the technological imperative becomes a political imperative: federalism bends and energy-policy needs proceed on their way. The constraint is only that; it is not a prohibition to felt needs of the energy issue.

References

Aron, Joan B. 1979. "Intergovernmental Politics of Energy." *Policy Analysis* 5 (Fall): 451–471.

California Energy Commission. 1981a. *Energy Tomorrow: Challenges and Opportunities for California.* 1981 Biennial Report.

_____. 1981b. "Moving toward Security: Strategies for Reducing California's Vulnerability to Energy Shortages." Draft report.

_____. 1981c. "Moving toward Security: Strategies for Reducing California's Vulnerability to Energy Shortages; Appendices." Draft report.

Colglazier, E. William. 1982. *The Politics of Nuclear Waste.* New York: Pergamon Press.

Energy Impact Assistance Steering Group. 1978. *Energy Impact Assistance.* Report to the President. Washington, D.C.: Department of Energy.

Depree, A. Hunter. 1957. *Science in the Federal Government: A History of Policies and Activities to 1940.* New York: Harper and Row.

Dubnick, Melvin J. 1982. "The Delivery of National Policy Mandates." Paper delivered at the 1982 ASPA National Conference on Public Administration, Honolulu, March 21–25.

Dubnick, Melvin J., and Gitelson, Alan. 1981. "Nationalizing State Policies." In Jerome J. Hanus, ed., *The Nationalization of State Government. Lexington, Mass.: Lexington Books, D.C. Heath and Company.*

Goodwin, Craufurd D., ed. 1981. Energy Policy in Perspective. Washington, D.C.: Brookings Institution.

Grodzins, Morton. 1966. *The American System.* Edited by Daniel J. Elazar. Chicago: Rand McNally.

Hall, Timothy A.; White, Irwin L.; and Ballard, Steven C. 1978. "Western States and National Energy Policy." *American Behavioral Scientist* 22 (November–December): 191–212.

Kearney, Richard C., and Garey, Robert B. 1982. "American Federalism and the Management of Radioactive Wastes." *Public Administration Review* 42 (January–February): 14–24.

Lee, Kai N. 1981. "Regional Power and Local Governments." *Washington Public Policy Notes.* University of Washington, Volume 9.

National Energy Policy Plan. 1981. "Securing America's Future." Washington, D.C.: U.S. Department of Energy.

Pelham, Ann, *Energy Policy.* 2d ed. Washington, D.C.: Congressional Quarterly.

Plummer, James L. 1977. "The Federal Role in Rocky Mountain Energy Development." *National Resources Journal* 17 (April): 241–260.

Reagan, Michael D. 1969. *Science and the Federal Patron.* New York: Oxford University Press.

_____. 1982. "Energy: Government Policy or Market Result?" Paper prepared for delivery at the 1982 American Political Science Association meeting, Denver, September 2–5.

_____ and Sanzone, John G. 1981. *The New Federalism.* 2d ed. New York: Oxford University Press.

Regens, James L. 1980. "State Policy Responses to the Energy Issue: An Analysis of Innovation." *Social Science Quarterly* 61 (June): 44–55.

Shapiro, Fred C. 1981. *Radwaste.* New York: Random House.

Spangler, Miller B. 1980. Office of Nuclear Reactor Regulation. *Federal-State Cooperation in Nuclear Power Plant Licensing.* Report NU REG: 0398. Washington, D.C.: U.S. Nuclear Regulatory Commission.

Wilbanks, Thomas J. 1981. "Local Energy Initiatives and Consensus in Energy Policy." Paper prepared for the Committee on Behavioral and Social Aspects of Energy Consumption and Production, Oak Ridge, Tenn.: Energy Division, Oak Ridge National Laboratory.

2

The Impact of Market Structure and Economic Concentration on the Diffusion of Alternative Technologies: The Photovoltaics Case

Thomas Dietz and
James P. Hawley

Social scientists have studied the process of technological innovation for decades. Social theorists have considered the effects of technological change on the labor process, the nature of work, life-styles, and a wide range of other elements of social life. A large literature has developed on the adoption of new inventions by users and on the resulting diffusion of innovation through a pool of potential users (Radnor, Irwin, and Rogers 1978). A number of economists and management scientists have examined the process by which innovation, or invention, occurs within a firm or other organization. Implicit in most of this research is a process model of technological change: a new technology (or modification of an existing technology) is developed in an institutional setting, usually the research and development (R&D) division of a large or medium-sized corporation or a small firm. The innovation is marketed and then adopted by consumers. Finally, social changes of various sorts occur as the new technology is put into use by adopters. This model assumes that the transition from research laboratory to the market is a direct and simple step, based on calculations of the potential profitability of the innovation. The purpose of this chapter is to argue that the translation of invention into marketed innovation is in fact a complex process that requires careful study in any analysis of innovation.

The Process of Innovation

For purposes of discussion, the process of innovation can be broken into four steps: invention, implementation, marketing, and adoption. The distinctions among these steps are somewhat arbitrary, but they are analytically useful and correspond to most distinctions in the literature. Figure 2–1 illustrates the linkages between the steps.

Phase	I: Supply-Side Invention/Innovation[a]	II: Supply-Side Implementation of Innovation: Commercialization	III: Demand-Side Diffusion/Adoption[b]	IV: Supply-Side Market Strategy
Topics studied	Sources of invention Individual genius R&D patents—output $ spent on inputs social organization of R&D	Rate of introduction of products/processes Price decisions Product mix Government policies, supply subsidies	Nature of markets and communication networks Impact of large purchases on market development Catalyst to implementation and innovation	Market research; consumer behavior

[a]Internal to the firm—a more or less controlled environment; nonmarket principles.
[b]External to the firm, relatively less controlled environment; market or quasi-market principles.
Note: Arrows indicate feedback used to influence decisions.

Figure 2.1 The Phases of Innovation

Invention

Invention here means the development of a new technology or of a modification of an existing technology from the point of an initial idea through a working prototype to the point of pilot plant production. Invention is the responsibility of the R&D division of the modern firm. A number of theorists have looked at the relationship of firm size, market structure, and invention. Schumpeter (1942) and Galbraith (1956, 1978) have argued that large firms with oligopolistic control over markets are the natural home for most invention in contemporary capitalist societies. It is certainly true that large, oligopolistic firms have more resources to devote to R&D activities and are more likely to have a large enough R&D operation to maintain state-of-the-art equipment, attract top personnel, and maintain a critical mass (Scherer 1965). But the large size of these firms also imposes a bureaucratic structure on them, which may be antithetical to creativity. The overall direction of R&D in the large firm is usually strongly influenced, if not completely controlled, by the strategies of the marketing division and top management rather than by an understanding of the most-fruitful targets for research (Blair 1972, pp. 228–257). The semiconductor industry (especially in its early stages during the late 1960s and early 1970s) is often cited as the prototype of an innovative industry, yet most of the firms in this industry were of intermediate size during the period when most new technologies were developed. The majority of studies that relate concentration and scale of firm to invention conclude that an intermediate-scale firm in a somewhat but not intensely competitive market tends to be most efficient at producing new technologies (Kamien and Schwartz 1978; Mansfield 1963; Mansfield et al. 1971, pp. 172-174, 222-230; Parker 1974, pp. 60-75; Swan 1970, pp. 627-630).

Implementation

The invention stage of the technological innovation process ends with the development of a working prototype. The next stage is implementation, converting the inventions into a marketed product. Much conventional analysis ignores this stage, implicitly or explicitly assuming that a straightforward calculation of potential sales, prices, and production costs determines whether a firm will market an innovation and that once a decision is made, the new technology is turned over to marketing specialists. In constrast, we believe that the decision to develop prototype facilities and the choice of marketing strategies are the critical steps in the innovation process, and no clear understanding of innovation can occur without detailed knowledge of the implementation process. Practitioners and management analysts have long recognized the importance of the implementation process but typically have restricted their attention to the firm level. They have consequently ignored the relationship of the implementation process to industry structure and market power and share.

While little research has been done on implementation, there is substantial evidence that it is a key step. Parker (1974, p. 40) compares two estimates of costs incurred to bring an invention to the market. The estimates differ significantly on some costs, but both estimates indicate that prototypes or pilot plant and tooling and manufacturing start-up dominate the costs of bringing an invention to market. It is this transition from invention to marketing that places the greatest drain on a firm's resources, is the riskiest step for the firm, and thus is the highest hurdle in the overall innovation process. The development process in most R&D studies has been slighted in favor of the research emphasis, although the former consume well over half of the total costs in bringing a product to market (Rosenberg 1976, pp. 76–77).

Marketing

The marketing of a new technology might be thought of as the supply-side complement to adoption by consumers. Many decisions about marketing strategy may be made during the implementation phase. Included in the marketing phase are only the actual efforts of the firm to promote its product, not the earliest evaluations of market size and characteristics, which are central to the decision to move from the prototype technology to production. Marketing is intimately linked to adoption. Both processes take place simultaneously and influence one another. Monitoring of sales can lead to shifts in marketing strategy; changes in strategy will change the pace of adoption and the groups who are adopting. The rationale for making an analytical distinction between adoption and marketing is to emphasize that diffusion should not be thought of as a process with its own internal dynamics but rather as the cumulation of adopter decisions where those decisions are strongly influenced by advertising and promotional strategies. These strategies in turn are shaped by the structure of the industry and by various tactical and strategic decisions of the firm, and divisions of those firms, in the industry. There is substantial evidence that in certain instances, oligopolistic firms have withheld from implementation and marketing technology that would have undercut its previous investments or alternate product lines. (For instance, see Abernathy 1978, pp. 35–38, 55–60; Blair 1972, pp. 213–245; Gilbert and Newberry 1979, pp. 1–10; and Jewkes, Sawers, and Stillerman 1959, pp. 177–187, 215–222).

Adoption

Adoption is the demand side of the diffusion process. The decision of individuals or organizations to adopt a new technology is not wholly dependent on the internal characteristics of potential adopters and their strategies but is also heavily influenced by marketing strategies of the firms purveying the new technologies. In the case of new technologies of military importance, such as jet engines, semiconductors, and microprocessors, government policies may play a critical role in demand stimulation. The relationship between the firm's marketing strategy and government procurement policy is consequently important.

We have described the process of innovation in a linear fashion. Obviously this is a simplification. At every stage of the process, managers within the corporation are looking at other stages, trying to anticipate further inventions, implementation strategies, and marketing tactics. Decisions at each stage are made on the basis of expectations about the behavior of competitors as well as internal firm developments (White 1981). These complex managerial and bureaucratic decisions are not easily understood by application of neoclassical, microeconomic theory. As Mansfield (1968, p. 486) has noted, "With regard to many of the major issues concerning basic research, economics has little to say. As one works towards the development end of the R&D spectrum, economics becomes more useful, but it still has a limited contribution to make." The problem with the neoclassical approach is that it ignores the internal dynamics of the firm. Abernathy (1978, p. ix) in his study of innovation and productivity in the U.S. automobile industry notes that "innovations shape the course of industrial process but [are] in turn directed and then subdued by the competitive pressures *within* the firm [emphasis added]." It is this aspect of the innovation process that most researchers have ignored. Abernathy emphasizes the trade-off between innovation in products and production processes and increased productivity and argues that "few books have considered how innovation does or does not fit into the mainstream of a corporation's competitive plans, why some firms seem to emphasize innovation in their competitive actions whereas others rely more on marketing or financial or price competition."

The Decision to Implement

The factors that impinge on the decision to develop and market an invention can be divided into two general categories: organizational factors, the internal politics and structures of the firm itself, and market factors, both the market position of the firm and the structure of the market. Market position refers to the role and relative strength of the firm in the market in which the new technology would compete. Market structure refers to the overall structure of the market. Obviously each of these sets of factors conditions the others, but the distinctions are useful because they emphasize the differences between interorganizational and intraorganizational influences on the innovation process.

Organizational factors

The key organizational factors influencing the implementation of innovation are the size, resources, and structure of the firm. Larger firms have more financial and technical resources available to support development of inventions into products. But large firms are also more bureaucratic than smaller firms, so expertise in various areas must be integrated by committee rather than in the mind of the manager (Freeman 1974, pp. 109–110). Communications become formalized, so marketing and manufacturing experts may have little contact with research staff and may have only superficial understanding of a new invention. Increasing size of organization is usually accompanied by diversity in product lines and activity in multiple industries. Diversified operations allow profits from some markets to cross-subsidize innovation in markets where risks or costs might not otherwise be acceptable. But these other divisions are also competitors for scarce resources. Indeed the marketing of a new technology may cut into the sales of an existing technology. In this situation, a new technology may be profitable on its own but may be withheld from the market because it could interfere with sales of an existing product line. In addition to these formal considerations, organizational politics may play a critical role in the decision to implement. The marketing of a new technology may require the creation of new organizational units within the firm or the reallocation of resources across existing units, and this may lead to strong support for or opposition to a new technology regardless of its intrinsic merits.

These and related research questions can be fruitfully investigated less by received microeconomic analysis and assumptions geared toward the analysis of markets in decision making than by an institutionally informed focus on hierarchies, organizational politics, strategies, and structure (Lindblom 1977, pp. 84–85; White 1981; Williamson 1975, p. xi). A purely market-oriented analysis assumes some form of exchange among autonomous entities, decision making being informed by ultimate consumer markets. A focus on hierarchical and organizational politics looks toward corporate structure as a single administrative entity, with sub-ordination among its elements as a dominant characteristic. The typical corporate structure in this sense is corespective, more concerned about its few and knowable competitors than with autonomous and ultimately unknowable consumers (White 1981). Decision making in most corporate hierarchies is not with the single-product, single-industry perspective that microeconomic analysis assumes. Rather a diversified firm confronts numerous potential markets, product lines, and possible strategies.

One manifestation of the narrowness of microeconomic analysis is pointed out by Miller (1971, pp. 11–12) and Abernathy (1978, p. 168). Miller suggests that most economists have defined innovation only with regard to its newness, neglecting, for instance, imitation of a prior innovation by competitive firms as inconsequential. Innovation viewed as an organizational process, on the other hand, does not define innovation a priori but in terms of what is of actual importance to a firm's strategy and structure. Abernathy's study of innovation and productivity in the automobile industry makes a similar point. Innovation and productivity are typically trade-offs, which a firm's management must consider in its competitive and organizational strategies.

Most studies of innovation pay scant attention to the industrial structure of the firm. Thus, a widely cited study of steel-industry innovation does not concern itself with a particular firm's strategic planning aside from innovation, such as increased vertical integration or diversification (Mansfield 1968, pp. 48, 68). In contrast, Abernathy (1978, p. 168) suggests that this may be a central issue. He defines the conceptual object of his study as the productive unit, which combines the product and the characteristics of manufacturing process that produce it. This productive unit cannot respond to all of the demands made upon it; it rarely can be maximally both efficient (productive) and simultaneously innovative. Thus corporate management must "manage a portfolio of productive units."

A few studies attempt to analyze this management decision process involving a portfolio of productive units. Fewer still take into consideration the several or several dozen productive units (business lines and means of producing those lines) at various stages of development. Yet the typical unit of production is the diversified, if not conglomerate, corporation that manages just such units in large multiples.

Abernathy's (1978) initial evidence suggests that firms effective at one stage of innovation are less effective at others. Auto, steel, and oil firms are cited as examples of efficient mass producers and infrequent "sources of radical new products." In contrast, for instance, are the high-technology electronic firms, which often have difficulty competing in high-volume markets.

Market Factors

The relative strength of a firm in a market has an important influence on decisions to introduce new technologies. A firm that dominates a market may be in a better position to take the risks of introducing a new technology, but it may have less incentive to do so. It is important to distinguish between a firm's share of a market and the overall structure of that market. A firm with a large share of a competitive market will face the risk that a competitor innovating or obtaining a new technology will cut into existing sales of current technologies. This may be an incentive to implement a new technology, or it may argue for developing the new technology to the point where it can be controlled, even if not implemented. This latter strategy may be achieved by ownership of key patents, by buying out new entrants into the market, or by pricing existing technologies so that new technologies cannot be sold in large enough quantities to justify the costs of their development or obtain production economies of scale. Thus dominance in a competitive market offers both incentives and discincentives to innovation. In contrast, it seems likely that the absence of competition and of diversity in a market retards innovation (Mansfield et al., 1971, p. 13). Profitable oligopolistic firms have little to gain from replacing existing technologies since they can effectively block new entrants and since new technologies still carry significant risks. Thus there are few incentives and a strong discincentive for innovation in highly concentrated markets with stable and predictable demand structures.

This outline of factors influencing the implementation of new technologies is not intended as an explicit theory of implementation. We simply want to emphasize the importance of a variety of intraorganizational and interorganizational factors by indicating a few of the complexities involved in the decision to implement that are not treated in the classic literature on the subject. A theory of implementation must wait until further exploratory work can be completed. We have begun such an analysis of the solar photovoltaic industry. The next section presents initial results from this work.

Implementing Innovation in the Photovoltaic Industry

Technology Profile

In order to analyze the implementation of innovation, a profile of photovoltaic technology and production techniques is essential. The photovoltaic process converts sunlight into electricity quietly and without mechanical movement. When the photons in a beam of light strike an atom (for instance of silicon), they interact, and the additional energy can drive one of the outer electrons off the atom. Different photovoltaic material (for example, silicon and cadmium sulfide) absorb photons at various wavelengths producing electricity. Silicon and other semiconductor material are effective photovoltaic converters because most of the entering protons will disperse electrons up to a maximum theoretical efficiency, for silicon of 23 percent. The remaining energy is transformed into excess heat (Maycock and Stirewalt 1981, pp. 17–27).

Photovoltaic Cell Material

Three types of silicon photovoltaic cells are currently in production and the advanced R&D stages in the United States. Single-crystal silicon is the most commonly produced commercial type of silicon. It has relatively high conversion efficiencies (8 to 15 percent) and has a long, nondegrading active life of between twenty and thirty years. It is, however, expensive to produce, resulting in a high final price of electricity (measured in watts at peak insolation—Wp). A second type of silicon, much less expensive to produce, is polycrystalline silicon. Because single-crystal silicon must be "grown" in a vatlike machine under extremely complex conditions and taking substantial time, it is expensive. Polycrystalline silicon, on the other hand, can be produced in a variety of forms—ribbons, ingots of various shapes, blocks—and can also be directly deposited on a flat surface, or substrate. This greatly reduces costs compared with single-crystal silicon. Polysilicon, however, has a lower conversion efficiency than does single-crystal silicon because tiny electrical short circuits take place along the boundaries of its many crystals. Also the poly-cell degrades more quickly than does the single cell.

The third type of silicon is amorphous silicon. It is a pure silicon with no crystal structure, its atoms randomly distributed. This has a tremendous cost-reduction potential since only small amounts of silicon are used in the production process, and there is no need for the costly cooling and growing processes to create either one or many crystal structures. The trade-off, however, is that amorphous silicon's efficiency is low, currently only between 3 to 6 percent. (Efficiency is directly related to the module size necessary to produce a given amount of electricity. The lower the efficiency, the larger the module and array size required to produce a given Wp.) It is possible that significant technical breakthroughs will occur, making amorphous production more than competitive with single-crystal silicon.

Photovoltaic R&D using nonsilicon materials is proliferating. There are dozens of materials and combinations of materials possible to produce photovoltaic efficiencies up to 27 percent (Maycock and Stirewalt 1981, p. 37; California Energy Commission 1981, pp. 79–134). The two currently most-promising nonsilicon materials are cadmium sulfide and gallimum arsenide cells, yet there are many technological and economic drawbacks at this point, which focus primarily on the economics of production and process technology.

Photovoltaic Production Processes

The first major current expense in silicon-based photovoltaic production is the production of an extremely pure grade of silicon containing less than one impure atom per billion. This resulted in a 1980 selling price for semiconductor silicon of $50 to $70 per kilogram, or about half of the total cost of a finished cell. Immediately promising cost reductions are on the horizon, which can provide significantly lower-cost silicon and photovoltaic processes, which can tolerate less pure silicon, perhaps in the $10 to $20 kilogram range.

Single-Crystal Silicon: First-Generation Technology: Purified polycrystalline silicon is remelted in a vat of molten silicon in which a "seed" of silicon on a holder is dipped into the melt and then pulled upward at the rate of about an inch each three hours. A perfect crystal structure can thus be "grown" as the silicon solidifies around the seed. Ingots of up to about six inches can be pulled in this fashion. Technological improvements in the speed of this process have been incremental, with severe future limits. This process represents the first generation of photovoltaic single-crystal silicon production.

After an ingot is grown, it is sawed into wafers, with silicon losses of about 50 percent. Thus the purity of the original silicon, the slowness of the growing method, and the large losses from sawing result in a very high cost per cell.

While still molten, the silicon is doped with either boron or gallium to produce a positive electronic charge. After the wafer is cut, it is coated on one side with phosphorus (or some other element) to produce a negative charge. Finally, metal contacts are fused onto the cell on each side and the whole cell encapsulated in a coating such as plastic or glass.[1]

Silicon Sheet Production: Second Generation: Avoiding the pulling or growing and sawing process would reduce cell costs significantly. Two methods of sheet production are under intensive research and development and in two cases full production: ribbon production and the dendritic web processes. Both have the potential for continuous-process production of high-quality and -efficiency silicon, at significantly lower costs. These processes are second-generation, single-crystal silicon. Commercial production using these methods began in 1981.

A third, still-experimental cost-reducing method of silicon production is to make single-crystal material in an ingot casting (rather than a growing) process and then slice it. The casting methods can also be used to produce polysilicon, which requires less-pure and thus less-expensive silicon.

Another process (not currently in commercial production) places a thin coat of silicon on a ceramic substrate, which continuously moves along an assembly line. Ceramic thin film thus has the potential for continuous production, resulting in a high output to capital ratio, and producing a product ready to mount in a module. It uses only about 25 percent of the silicon of other processes, further reducing costs. Its quality and efficiency remain to be demonstrated, however.

Amorphous Silicon Production: Third Generation: This is the third generation of photovoltaic technology. It is inherently less expensive than other silicon processes and types and is being improved. The production process bypasses crystallization processes, uses less silicon, and is durable. If efficiency can be raised, it would be a significant technological innovation with immediate commercial importance. It is currently commercially produced for gadgets only by Japanese firms.

Nonsilicon Thin Films: These are similar in production technique to silicon thin films but have the advantage of lower material costs. Also some of these materials can be coated directly onto glass and processed at lower temperatures than silicon requires. Pilot production involving a number of materials (including cadmium sulfide and copper sulfite) are in production, but to date these materials tend to degrade rapidly once in use and have signiciantly lower cell efficiencies than silicon, thus requiring larger array areas and balance of photovoltaic system requirements, increasing price per Wp. Nevertheless, thin film may be a significant future competitor to silicon technology.[3]

In sum, single-crystal silicon cells grown in a vat is current first-generation technology. The products are efficient and reliable, but are still expensive. Second-generation production techniques include various methods of continuous generation of poly-crystalline silicon (such as ribbon, square ingot, and dendritic web), which will reduce production cost significantly, although currently with some loss of cell efficiency. In early 1982 commercial production using one of these second-generation methods began. The third generation is represented by R&D in thin film materials and in R&D and initial pilot production for extremely small-scale uses of amorphous silicon. A fourth generation is represented by using multiple layers of photovoltaic materials and light concentrators to achieve high or total efficiencies. Concentrators, however, can use various cell types while extending peak isolation periods longer throughout the day by making investment in tracking support structures more economic.

Technological innovation in the photovoltaic industry in the United States and worldwide is developing rapidly into commercial sales of second-generation technology. Most industry observers and participants indicate that innovation will continue to be slow, resulting in the maintenance of significant competition among cell types and balance of systems-support structures for U.S. and foreign producers.

Markets, Exports, and Global Competition

The U.S. photovoltaics industry is rapidly growing, dependent on foreign exports for between two-thirds and three-quarters of total production. There are essentially two markets. The first is for remove, stand-alone, off-electrical-grid applications. The second market, or potential market, is for photovoltaic installations as a complement to, or a substitute for, central-station electric generation.

The stand-alone market has two sectors. The first is for uses such as photovoltaic-powered communication equipment, navigational aids, railroad-signaling devices, and cathodic protection for oil wells and bridges. The second sector of the stand-alone market is for water pumping and off-grid power equipment, especially important in Third World locations for refrigeration and local village power in general. In this latter market, the alternative source of electricity is typically diesel-powered generation, so photovoltaic market penetration is highly sensitive to the price of photovoltaics versus the price of competing fuel plus diesel maintenance costs. The village market has high potential but is limited by available foreign-exchange reserves of the various countries. Nevertheless, it is the most rapidly growing sector of the current photovoltaic market (U.S. DOE/NASA 1980:193).

In sum, the potential of these markets provides a basis for significant production runs and continued production expansion. Yet the rapidity with which end-use technology and production processes are developing affect some industry participants that have committed themselves to first-generation technology.

The second market is composed of grid-connected residential and industrial, and off-grid-connected utility and decentralized miniutility generation. There are not yet commercially competitive uses of photovoltaics, although the number of experimental and demonstration projects is increasing rapidly, subsidized by a variety of national and local state tax credits, as well as by government grants and as part of R&D activities by producers and utilities. This limited market can have significance for innovation in the industry as well as for its structure because utility and other large purchasers are often able to influence standards, types of technology, and consequently production economies of scale.

The nonsubsidized photovoltaic commercial market is primarily an export market since the majority of current commercial end-use applications are commercially suited to off-grid, stand-alone sites. This suggests that those firms most knowledgeable about and best connected to foreign, especially Third World, nations will fare better in the increasingly competitive global photovoltaics market. As prices continue to decline, the proportion of exports may be reduced as photovoltaics become commercially competitive with other continental U.S. energy sources. This, however, is closely related to federal and state policy, especially utility and tax policy, and is not likely to change rapidly throughout the 1980s.

The photovoltaic market is a segmented market; there is no single price at which photovoltaics become immediately and obviously competitive with other energy sources. Thus, targeted public-policy action can serve as an important catalyst to faster innovation as it serves to lower production and distribution costs through increased economies of scale. Government's power to purchase large quantities and to tax and set utility policy and rates will continue to be critical variables in the photovoltaics industry innovation process. This is particularly the case since European and Japanese photovoltaic firms enjoy substantial R&D and market support from their respective governments, although U.S. firms maintain a technological lead over foreign competition.

Photovoltaic Price Trends

The combined forces of a rapid rate of technological innovation with growing world markets (at over an 80 percent compound annual rate between 1978 and 1982) have resulted in significant price declines. The commercial applications of photovoltaics depend on the prices at which different systems can economically compete with alternate energy systems, taking into account any tax credits that may exist. The 1982 average selling price (ASP) for photovoltaic flat-plate modules is about $10 per Wp, although there are documented cases of one firm's price as low as $6.50 per Wp, while other firms offer comparable modules at up to $20 per Wp.

As table 2-1 indicates, the worldwide ASP has been reduced by almost half since 1977, declining from about $19 per Wp to $10.65 per Wp in 1981. Similarly worldwide shipments (kW) have grown from 450 in 1970 to 5,500 in 1981, a twelvefold increase, and are expected to double that in 1982. National photovoltaic industrial capacity between 1979 and 1982 greatly exceeded shipments by what is a common estimate of at least 50 percent. Thus, current U.S. photovoltaics capacity is much greater than are shipments. A number of industry observers have estimated that for 1982 the ASP worldwide will be $9 to $10 per Wp, with total shipments between 9,000 and 10,000 kWp, resulting in total global sales of about $90 million. It is variously estimated that U.S. production in 1981 accounted for at least half of global total sales and perhaps as high as two-thirds, although U.S. photovoltaic industrial capacity is probably well over half of global capacity.

Table 2-1
Global Photovoltaic Module Market Sales

	Shipments (kWp)	ASP (kWp)	Dollars (millions)
1977	450	19.00	8.6
1978	950	14.70	14.0
1979	1,450	13.50	19.6
1980	3,250	12.00	39.0
1981	5,500	10.65	50.0
1982[a]	9,000	9.00–10.00	90.0

Source: William J. Murray, "PV in 1980: Growth Accelerates," *Solar Engineering* (January 1981): 14, and interviews.
[a]Estimated.

ARCO Solar's recent agreement with Southern California Edison for a one m Wp generation station, the largest photovoltaic installation in the world, will account for about 10 percent of the estimated 1982 world market, a reflection of significant U.S. surplus industrial capacity. It is not yet (and may never be) known the actual price for this agreement since ARCO Solar will, in effect, sell photovoltaic modules to itself under the arrangement. Since significant tax credits based on selling cost will accrue to ARCO Solar, we can assume that the book selling cost will be relatively high. If true, this will significantly raise the world selling price for 1982 beyond what an actual market price would have been in absence of this arrangement.

In sum, significant price reductions can be expected throughout the 1980s as new production processes, economies of scale, learning-curve declines, significant global competition, and excess capacity push prices down. By the late 1980s, the global photovoltaic industry will be a billion-dollar industry.

Price declines and the growth of global competition combined with recent substantial surplus capacity has placed great financial pressure on all industry participants, almost none of whom have operated profitably since about 1978. This has resulted in a tendency toward increased economic concentration within the industry and toward industry domination by large diversified corporations entering it through buy-ups, joint ventures, and subsidiary establishment.

Ownership Structure and Economic Concentration

The photovoltaics industry has rapidly developed from a small industry in 1977 whose typical firm was small, founded and owned by an independent, entrepreneur scientist (in addition to two large firms producing for the space market only) to one that is dominated by subsidiaries or joint ventures of large, diversified global corporations (especially oil-based energy corporations), in the United States as well as in Europe. (Japanese firms tend to be those previously engaged in semiconductor manufacture.) In mid-1982 there were only four operating U.S. independent photovoltaic producers of approximately twenty producers. These four independents (one of which is primarily a capital-equipment producer) account for what we estimate to be less than 5 percent of total U.S. production. The photovoltaic industry is a large corporate industry.

The following list presents the ownership structure, by the patent firm's industrial sector, of the U.S. photovoltaic industry:

1. Oil company owned (in whole or in part)
    Arco Solar (100 percent ARCO)
    Mobil-Tyco Solar Energy Corporation (80 percent Mobil, 20 percent Tyco)

Photon Power (50 percent Total Oil, French)

Photowatt (Societe Français de Photopiles, majority owned by Elf Oil, and ultimately owned by the French government)

Solarex (25–40 percent by AMOCO)

Solarvolt (a joint venture between Motorola and SES, owned by Shell Oil)

Solar Power Corporation (100 percent Exxon)

2. Glass-based ownership

Applied Solar Energy Corporation (50–70 percent of publicly held shared owned by Optical Coating Laboratory, 30–50 percent by public)

Solec International (80 percent by Pilkington Brothers, Great Britain)

3. Independents

Richway Enterprises (not yet in operation)

Silicon Sensors

Solenergy

Sollos, Inc.

Spire Corporation (capital producers)

Energy Conversion Devices, an independent, has received $28.3 million from ARCO and about $80 million from SOHIO for research and development. [Stambler and Stambler, 1982:B 3–4; and interviews with industry executives]

Clearly the future of the U.S. photovoltaic industry innovation is in the hands of large, diversified, and primarily oil corporations. This trend has caused concern among some observers who have suggested that among other consequences such ownership patterns could retard innovation. It is also suggested that since the late 1970s through mid-1982, there has been an apparently high concentration ratio among U.S. producers. These related trends of high concentration and ownership by large, diversified corporations have caused some to suggest that the ability of oil and other relatively wealthy firms to support and buy up photovoltaic firms could have a significant number of detrimental consequences. (For instance, Reece 1979; Stambler and Stambler 1982; U.S. Congress, House 1980; U.S. Congress, Senate 1980). The most important among these are alleged to be: (1) the potential for technological suppression or retardation to protect heavy capital commitments to existing technologies; (2) the bureaucratization of the invention-innovation process within large firms; (3) the use of cross-subsidization to enable large, diversified parent companies to sell substantially below cost of production to gain market share, and consequently to raise artificially the perceived start-up costs for new industry elements; and (4) the ability of giant, and especially diversified oil-based, corporations to operate effectively and efficiently what is clearly a very different type of industry.

The first fact to stress is that the photovoltaic industry is primarily global. Measuring and attributing significance to U.S. market share is dependent on this fact. Market share is defined here to mean the proportion of total sales or of production by U.S. firms within the U.S. market; that is, total U.S. production is the denominator, and total number of firms' share of production is the numerator. This standard definition of market share is nationally bound, measuring production rather than market share. The domestic market share will be substantially larger, given increasingly serious international competition, than U.S. world market share, which currently and in the next five to ten years will account for the majority of total U.S. production. Very high concentration ratios have been found in the U.S. photovoltaic industry for the years between 1977 and 1981 (Ethridge 1980, p. 44; Stambler and Stambler 1982, pp. 10–14). Most concentration data are for 1979 and 1980, the years when European and to a lesser degree Japanese competition was increasing rapidly. In the future we should consequently expect U.S. production to be less important worldwide, although still remaining dominant at least through 1985. Thus a significant reduction in the recent past and current U.S. concentration levels when viewed from a global perspective are to be expected. We expect the global photovoltaic industry to remain highly competitive based on both price and technology in the foreseeable future, especially because no one firm or group of allied firms appears able to control or monopolize the expected diffusion of various technological innovations in second-, third-, and fourth-generation technologies.

Technology versus Market Strategy

In this context firms are pursuing different mixes between two possible polar firm strategies. The first strategy is what we call market strategy, pursued most aggressively by ARCO Solar, Solarex, and Solar Power Corporation. It places primary emphasis on market presence and market share, with significant capital and cost commitment to existing single-silicon-crystal technology and the rapid expansion of its production capacity, global marketing apparatus, and service network. While evidence is certainly lacking due to proprietary data, circumstantial indicators suggest that such a strategy, in the absence of continual and quite substantial R&D subsidization from the parent corporation—all three firms are assumed to be operating at a loss, as are almost all other photovoltaic firms—will of necessity pay less attention to the relative advantages of new technologies and process innovation, thereby retardng (and in the extreme, suppressing) implementation of new product and process innovations. This can be the case even if an active and fruitful R&D policy is maintained within the firm (as evidence suggests ARCO Solar and Solarex, for instance, are doing) because the capital costs for implementing such innovations may be prohibitively high given recent and expensive investment in current technologies.

Substantial market power and market share, traditional signs of oligopolistic practices, enable such a market-based strategy to pay off, although in a technologically unstable environment characterized by expected if not imminent innovation, such strategy is extremely risk prone. Abernathy (1978, p. ix, 168) has suggested that this situation creates what he terms a trade-off between continual innovation (requiring the destruction of existing capital investments) and increased productivity (requiring the increased or continual commitments to existing capital structures). He suggests that firms committed to continual high-volume production and high or rapidly expanding market shares and/or a semblance of oligopolistic market power will tend to emphasize productivity gains to lower costs and thus prices at the de facto expense of innovation. On the other hand, firms with less market share will tend to be the innovators, having less to lose.

While the domestic photovoltaic industry is still too small and too young and the global market rapidly developing to draw firm conclusions, there is sound reason to think that a market-productivity strategy, even with current high concentration rates, will encounter severe obstacles, especially due to the unstable state of technological developments and the increasing competition of global market and the current unprofitability of such firms.

The alternate approach to the market strategy we call the technology strategy, pursued most consistently in the U.S. firms such as Photovolt, Mobil-Tyco, Westinghouse, and Solec International. Some firms have not made heavy capital commitments to single-crystal silicon technology, thereby enabling them (if capital is available) to adopt different cell-production technologies as they become available. Solec International (with the strong R&D links to Pilkington) is one such example. The technology strategy stresses pilot production of newer technologies along with the building of market, sales, and service networks.

A case-by-case analysis would be necessary to establish how each photovoltaic firm subsidiary of a parent corporation or joint venture operates as what Abernathy terms a productive unit. This would entail access to proprietary information (which we did not have), such as how R&D is cross-subsidized by the parent firm, how product prices are determined (and what they actually are), especially by firms selling below ASP, and how much autonomy each subsidiary or joint venture has at different points in its decision-making process from the parent firm. Lacking direct access to such data, we surmise that what Abernathy calls a portfolio approach to decision making by a parent firm's upper management will tend to stress either a market or a technology strategy.

Public policy has the potential to affect the strategy mix of firm strategies in two important manners. First, large public-sector purchases (for example, the Sacramento Municipal Utility District's or other utility purchases) of various technologies can be stimulated. Second, large block purchases have the effective power to set quality standards, thereby affecting a firm's commitment to a particular technology.

In our interviews for this study and in various other analyses of the photovoltaic industry, much attention has focused on ARCO Solar and Solar Power Corporation as two of the largest firms pursuing a market strategy, often using below-cost pricing to gain market share. Stambler and Stambler (1982, pp. I, 10–14, B-11) suggest that there was a four-firm concentration ratio (meaning that the four largest firms have a market share of a given percentage) of 86 percent in 1978 and 1980, leaving 14 percent of the market to the other ten U.S. producers.[4] While there is reason to think 85 percent somewhat high (70 to 75 percent is a better estimate), the major problem with this ratio is that it uses U.S. production for primarily global markets as the basis for judgment, along with a time period during which rapid changes in the international competitive environment were occurring.

With one exception in our interviews with industry officials, no one expected that the current levels of concentration would continue to exist. Most participants—almost all affiliated with major global corporations—suggested that the industry would not continue to be dominated by two or three firms (such as ARCO Solar, Solarex, and Solar Power Corporation), although few expected to see fifteen or twenty U.S. firms remain during the next five years. In sum, most interviewed thought that the current high levels of concentration were expected to drive them down. Typically oligopolistic market control and stability resulting in either artificially high prices or technological retardation are extremely unusual in a cross-subsidized industry characterized by growing global competition, rapid global technological innovation, and low or negative profitability.

Central to the concentration controversy has been the concern that one or two large producers have at times, and perhaps very consistently, been selling substantially below production and distribution costs, cross-subsidized by oil revenues, in order to gain market share. During the past two years, it has been reported to us that a number of complaints have been filed with the Department of Justice, typically against ARCO Solar for selling at between $6.50 to $7.50 Wp, while the going industry low price was at least $9.50 and up to $15 Wp. (Interviews with industry executives and Stambler and Stambler 1982, pp. 14–18.) This alleged practice, according to interviews, continues to mid-1982. Combined with an industry operating at about 50 percent of existing capacity and few if any firms turning a profit, the price competition at present is intense. Smaller companies with single-crystal-silicon technology believe they have lost customers and government contracts to low-price sellers, although there is little question that low prices have, artificially, expanded photovoltaic markets. In turn, this has had a positive impact, enabling early innovative adopters to purchase larger numbers of photovoltaic products, but also a negative impact by creating the illusion that technology is actually commercially competitive when it is not. In turn, the existence of artificially low selling prices by some firms along with substantial R&D investment by, for instance, ARCO, gives the impression that industry entry costs may be higher than they actually are, consequently discouraging new industry entrants and potentially discouraging innovation. Typically, in our interviews it was suggested that $50 million to $100 million is the current necessary cost to start up a photovoltaic operation, although other estimates are substantially lower, on the order of $5 million to $20 million.

Two other related concerns have been expressed about large-firm, especially oil-company, participation in the photovoltaic industry. The first is based on the general consensus of academic studies that conclude that the optimum firm size and market structure to promote innovation is a moderate-sized firm (having adequate R&D resources) operating in diversified and somewhat but not too intensely competitive market (Kamien and Schwartz 1978; Mansfield 1963; Mansfield et al. 1971; Scherer 1970; Swan 1970). While the absolute concentration ratio in the national photovoltaic industry would suggest that such a situation does not currently characterize the industry, telescoping diverse generations of technology would suggest otherwise, allowing for substantially greater latitude of future action than current concentration ratios alone imply.

The second related often-expressed concern views the large, conglomerate corporation as bureaucratically inclined, often but not always incapable of innovating creatively, especially inclined to choose market-share and productivity increases over innovational, longer term strategies. This in particular would apply to oil-company diversification into photovoltaics and other nonoil (or energy) related industrial sectors. If oil profits continue to decline relative to the near-record levels of a few years ago, less cross-subsidization would be available to photovoltaic divisions from the parent corporations. This will leave photovoltaic subsidiaries increasingly on their own, if not turning a profit after a few years, would be a potential target for corporate financial officers looking to sell off a division. This was the experience, for instance, of Exxon in the solar thermal industry when it sold Daystar, as well as Exxon's current troubled experiences with other nonoil investments, such as office equipment and electric motors.[5] While prediction is not possible in this regard, the photovoltaic growth rate is not as rosy as many corporate and government analyses suggested in the late 1970s nor is oil wealth a bottomless pit.

Conclusion

Our purpose was to examine aspects of the process by which technological inventions are transformed into implemented innovations. This process is complex, suggesting that a firm's decision is strongly influenced by both the organizational dynamics of the firm and most importantly by the firm's perception of its position within a market structure. In this regard the photovoltaic case would tend to support Jewkes's, Sawers's, and Stillerman's (1959, p. 222) conclusion that firm size itself is less important than the pattern, structure, and levels of industrial and technological development of an industry in influencing the implementation of innovations.

Notes

1. U.S. firms producing commercial single-crystal silicon cells are: Applied Solar Energy Corporation, ARCO Solar, Motorola (Solarvolt, a joint partnership with SES, a Shell Oil subsidiary), Photowatt International, Silicon Sensors, Solarez, Solar Power Corporation, Solec International, Solenergy, Sollos, and Spire. Stambler and Stambler 1982: B-2 and interviews.

2. Mobil-Tyco Solar Energy Corporation and Semix (a division of Solarex). Stambler and Stambler 1982, p. B-2.

3. Photon Power and Solarvolt engaged in pilot production. Stambler and Stambler 1982, p. B-2 and interviews.
Interviews have been conducted for this study with representatives from fifteen firms and with industry analysts with the prior understanding that respondents would be assured of confidentiality and no data about a particular firm revealed based on interviews with that firm's executives.

4. Typically, a four-firm concentration ratio between 70 percent and 75 percent is considered high.

5. Widely cited troubles for oil company diversifications have been commonplace in the business press during the last two years. This was pointed out by a number of former oil-company executives now involved in independent photovoltaic firms. Oil companies' strategies differ substantially from one another. For instance, Shell, Phillips, Chevron, and others have pursued a technology rather than a market strategy.

References

Abernathy, William J. 1978. *The Productivity Dilemma.* Baltimore: Johns Hopkins University Press.

Blair, John M. 1972. *Economic Concentration.* New York: Harcourt Brace Jovanovich.

California Energy Commission. 1981. "Solar Electricity for the 1980s." Mimeographed. Sacramento. April.

Chandler, Alfred D., Jr. 1969. *Strategy and Structure.* Cambridge, Mass.: MIT Press.

Ethridge, Mark. 1980. *The U.S. Solar Energy Industry and the Role of Petroleum Firms.* Research Study 018. Washington, D.C.: American Petroleum Institute.

Freeman, Christopher. 1974. *The Economics of Innovation.* Harmondsworth and Baltimore: Penguin Books.

Galbraith, John K. 1956. *American Capitalism.* Boston: Houghton Mifflin

————. 1978. *The New Industrial State.* Boston: Houghton Mifflin.

Gilbert, Richard J., and Newberry, David M.G. 1979. "Pre-emptive Pattenting and the Persistence of Monopoly." *Economic Theory Discussion Paper 15.* University of Cambridge, England, March.

Harter, Arthur P., Jr. and Rubenstein, Albert H. 1970. "Market Penetration by New Innovations: The Technological Literature." *Technological Forecasting and Social Change* 11:197-221.

Harvey, Edward. 1968. "Technology and the Structure of Organization." *American Sociological Review* 33 (April): 247–259.

Jewkes, John; Sawers, David; and Stillerman, Richard. 1959. *The Sources of Invention.* London: Macmillan; New York: St. Martin's Press.

Kamien, Mortin I., and Schwartz, Nancy L. 1978. "Potential Rivalry, Monopoly Profits, and the Pace of Inventive Activity." *Review of Economic Studies* 45 (October): 547–557.

Katz, Elihu; Levin, Martin L.; and Hamilton, Herbert. 1963. "Traditions of Research on the Diffusion of Innovation." *American Sociological Review* 28 (April): 237–252.

Lindblom, Charles E. 1977. *Politics and Markets.* New York: Basic Books.

Mansfield, Edwin. 1963. "Size of Firm, Market Structure, and Innovation." *Journal of Political Economy* 71 (December): 556–574.

————. 1968. *The Economics of Technological Change.* New York: W.W. Norton.

Mansfield, Edwin; Rappoport, John; Schnee, Jerome; Wagner, Samuel; and Hamberger, Michael. 1971. *Research and Innovation in the Modern Corporation.* New York: W.W. Norton.

Maycock, Paul D., and Stirewalt, Edward N. 1981. *Photovoltaics.* Andover, Mass.: Brick House.

Miles, Raymond E., and Snow, Charles C. 1978. *Organizational Strategy, Strategy and Process.* New York: McGraw-Hill.

Miller, Emile Roger. 1971. *Innovation, Organization, and Environment.* Institue de recherche et de perfectionement en administration, universite de Sherbroke, Quebec, Canada.

Parker, J.E.S. 1974. *The Economics of Innovation.* London: Longman.

Radnor, Michael; Fuller, Irwin; and Rogers, Everett M. 1978. *The Diffusion of Innovation: An Assessment.* Evanston, Ill.: Northwestern University.

Reece, Ray. 1979. *The Sun Betrayed.* Boston: South End Press.

Rosenberg, Nathan. 1976. *Perspectives on Technology.* London, England and New York: Cambridge University Press.

Rothberg, Robert R., ed. 1976. *Corporate Strategy and Product Innovation.* New York: Free Press.

Scherer, F.M. 1965. "Firm Size, Market Structure, Opportunity, and the Output of Patented Inventions." *American Economic Review 55* (December): 1097–1125.

_____. 1970. *Industrial Market Structure and Economic Performance.* Chicago: Rand McNally.

Schumpeter, Joseph. 1942. *Capitalism, Socialism, and Democracy.* New York and London: Harper and Brothers.

Stambler, Barrett, and Stambler, Lyndon. 1982. "Competition in the Photovoltaic Industry: A Question of Balance." Mimeographed. Center for Renewable Resources for Small Business Administration, Washington, D.C.

Swan, Peter L. 1970. "Market Structures and Technological Progress: The Influence of Monopoly on Product Innovation." *Quarterly Journal of Economics* 84 (November): 627–638.

U.S. Congress. House of Representatives. Subcommittee on Energy, Environment Safety, and Research of the Committee on Small Business. 1980. *Hearings: Role of Government Funding on Its Impact on Small Business in the Solar Industry,* 96th Congress.

U.S. Congress. Senate. Select Committee on Small Business. 1980. *Hearings: The Structure of the Solar Industry,* 96th Congress.

U.S. Department of Energy and National Aeronautics and Space Administration. 1980. "Market Definition Study of Photovoltaic Power for Remote Villages in Developing Countries." 0049-80/2, Washington, D.C.

White, Harrison C. 1981. "Where Do Markets Come From?" *American Journal of Sociology* 87 (November): 514–547.

Williamson, Oliver E. 1975. *Markets and Hierarchies.* New York: Free Press.

3

An Economic Analysis of Oil- and Gas-Leasing Experience under Profit-Share and Bonus Bidding with a Fixed Royalty

Walter J. Mead and
Gregory G. Pickett

Increased attention has recently been given to alternative systems for leasing publicly owned oil and gas resources. The U.S. Congress enacted the Outer Continental Shelf Lands Act Amendments of 1978 in which it mandated that bidding systems other than the conventional cash bonus bid with a fixed royalty must be used for not less than 20 or more than 60 percent of the total area offered for leasing each year during the first five years following enactment of the legislation. Bidding was authorized on any of four bid variables: cash bonus, royalty, net profit share, and work commitment. Each bid variable could be paired with any of the others as a fixed payment. This mandated feature of the act expires in 1983, and the issue of leasing alternatives must again be debated and resolved. Prior legislation authorized only cash bonus or royalty bidding. In practice, the interior secretary elected to require cash bonus bidding for 3,124 out of 3,162 outer continental shelf (OCS) tracts leased through the year 1977. Only 38 tracts (1.2 percent) were leased using pure royalty bidding.

Following the lead of the federal government, the state of Alaska passed legislation in 1978 in which it also required the use of bidding systems other than the conventional bonus bid with a fixed royalty. Specifically, Alaska required the use of profit-share bidding. While California has not yet reviewed its leasing statute, there is interest among legislators in expanding the authorized systems to include profit-share bidding.

This chapter evaluates two of the alternative bidding systems: cash bonus with a fixed royalty and profit-share bidding.

Bonus Bidding

Bonus-Bidding Procedures Followed by the U.S. Government

First among the economically meaningful leasing steps taken by the federal government is to announce a geographical area of interest. Potential bidders interested in the area are invited to nominate tracts for lease consideration by the government within the announced area. The government then specifies the tracts to be offered for competitive bidding. The list may include some or all of the tracts nominated and may include additional tracts that received no nomination. A time and place for the auction is identified. After the required environmental impact statements have been filed and the usual legal challenges have run their course, the auction is held, the bids are opened, and the results made public. The government then exercises its right to accept or refuse any and all bids tendered. The high bidder for each tract is declared to be the winning bidder.

The Economic Objective of Leasing Policy: Economic Rent Maximization

Congress has identified several, and therefore conflicting, objectives of its leasing program. From an economic as opposed to a political perspective, most economists would probably agree with McDonald (1979, p. 24) that government "should lease lands for minerals production on such terms and conditions, and at such a rate, as will tend to maximize the present value of the pure economic rent derivable from them." The term *pure economic rent* is defined as "the income which tends to accrue in the long run, under conditions of perfect competition and the absence of externalities, to the owners of land" (p. 26). Effectively competitive markets and absence of significant externalities are important requirements in these definitions. We will evaluate the effectiveness of competition in the market for OCS oil and gas leases in the next section. Whether significant externalities exist is beyond our scope here. However, we point out that federal and state government regulations have been issued designed to reduce to insignificance the probability of oil spills, oil or gas well blowouts, and other sources of environmental pollution, which constitute the principal external cost of OCS oil or gas production. Some external costs may still exist; however, they may be offset by external benefits in the form of national security or technological spillover benefits.

The concept of economic rent is easily illustrated in figure 3-1. The height of the bar represents the total revenue obtainable, subject to a cost constraint, from any given oil or gas lease. This gross revenue is reduced by the necessary costs of exploration and production shown in the lower segment of the bar. Under conditions of effective competition, the operator is forced to avoid unnecessary costs over which he or she exercises control. Government regulations that impose on lease operators social costs in excess of social benefits should be avoided. Such constraints either reduce production and revenue (lower the height of the bar) or impose unnecessary costs on the operator. In either case, the residual economic rent is reduced and society suffers a loss.

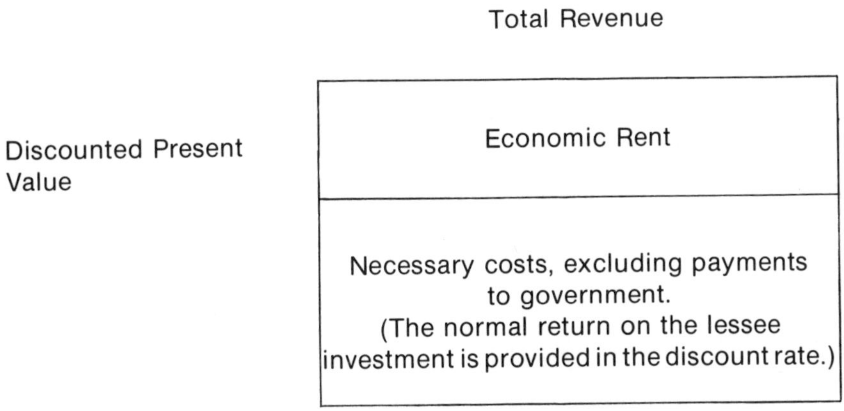

Total Revenue

Discounted Present Value	Economic Rent
	Necessary costs, excluding payments to government. (The normal return on the lessee investment is provided in the discount rate.)

Figure 3-1. Model of Economic Rent Estimation

Both the future revenue stream and all necessary costs are represented in figure 3-1 to be in present value terms. This means that a discount rate is used to compute present values. From a social perspective, the discount rate should reflect the marginal product of capital. Given the assumptions regarding competition and externalities, this rate corresponds with the discount rate that a firm would use in computing its optimal bid value. The discount rate includes a normal competitive rate of return on the capital employed.

Economic rent is paid in several forms under bonus bidding as practiced by the federal government. First, specified annual rent payments are required. These payments cease when production begins. Second, royalty payments amounting to one-sixth of the wellhead value of any oil and gas produced are required. Third, the discounted present value of the remaining economic rent is theoretically collected in the form of the cash bonus bid.

The leasing method used affects the amount of economic rent available for collection because the leasing method may affect the future revenue stream and the amount of the exploration and development costs imposed on the lessee.

Theoretical Evaluation of Bonus Bidding

The overriding advantage of bonus bidding as a means of determining to whom the lease should be awarded and for what consideration is that it harmonizes private incentives to minimize costs relative to any revenue stream, with the social-welfare goal of maximizing the economic rent. The economic rent is a residual. It is a transfer payment from the lessee to the lessor and is not a real social cost. Accordingly payments of economic rents in whatever form should not be allowed to affect operating decisions relative to whether additional oil or gas resources should be recovered. Under pure bonus bidding, the transfer of economic rent is accomplished by bonus payments. Once this payment has been made, the bonus becomes a sunk cost in future decision making. Any resources discovered on a lease become the property of the lessee. In pursuit of his self-interest, the lessee has an incentive to recover any oil or gas that produces incremental revenue in excess of incremental cost. Absent significant net externalities and given effectively competitive markets, this private incentive is in complete harmony with the general welfare. This means that the important operating decisions relative to whether to operate a well on which some amount of oil or gas has been discovered, or to plug and abandon the well, whether to make additional investments for secondary or tertiary recovery, and the like, and when to abandon a well in the final stages of its productive life are all made without the burden on economic rent payments included in marginal costs. In contrast, leases issued on the basis of royalty bidding create a divergence between private and social objectives. For example, the presence of royalty payments leads to premature abandonment of socially productive resources. To the extent that bonus leasing procedures followed by the federal government include a fixed royalty payment, this divergence occurs.

Profit-share bidding suffers from the same faults pointed out for royalty payments, but to a lesser degree. Depending on how the term *profit* is defined, there may be no payment to the government as a function of each unit of production (no marginal cost). However, because profit must be shared to a degree specified in the bid, the incentive for a lessee to produce a socially optimal reservoir and make additional optimal investment in order to increase production will be reduced and perhaps eliminated.

A secondary advantage of bonus bidding is that administrative costs are minimized. In a pure bonus-bidding system, the economic rent is paid in one lump sum. There is no need for the government to monitor investments in lease development. The lessee incentives and social incentives are in harmony. Further, there is no need to police or monitor production from the lease. Where a fixed royalty payment is appended to bonus bidding, some wasteful administrative costs are required. In any event, such costs are minor relative to the administrative burden on both lessor and lessee in monitoring profit-share payments.

The commonly listed disadvantages of bonus bidding are (1) by requiring a front-end payment of part of the economic rent, a barrier to entry is created that might have the effect of reducing competition for leases, and (2) there is no clear relationship between the amount of the bonus and the present value of the ultimate production, if any, from the lease. The former issue relates to the effectiveness of competition and is an empirical question. The latter issue is usually presented as a fairness issue relative to an individual lease.

Bonus Bidding as an Option Contract: Bonus bidding confers upon the winning bidder the exclusive right to explore, develop, and produce oil and gas from a specified tract within a given time period at an agreed-upon price. Such a contract bestowing the right to execute a stipulated transaction is termed an option contract (Smith 1976). This is in sharp contrast to a futures or forward contract is which the holder is obligated to carry out the specified transaction. This distinction is important because the payoff characteristics of these two types of contracts are quite different.

Only if the lessee were obligated to explore and develop the tract would bonus bidding be analogous to bidding for a futures contract. In general, however, lessees are free to terminate exploration and development activities. For this reason, the option to explore and develop the lease is a valuable asset. As long as there is some chance that the net present value (NPV) of the lease will be positive, firms will pay for the right to explore and develop the lease. That is, the lessee will never have to be paid to accept the right to explore and develop the lease. Therefore the winning bonus bid will always be positive. Although this result is relatively straightforward, mathematical models of bidding systems often do not yield this result as a conclusion. Instead these often specify that the lease will remain unsold if no positive bids are submitted. Clearly a model of rational bonus bidding must be based on behavioral postulates that yield positive bids as an implication. It is not enough to say that "negative bids are not allowed" (Reece 1979, p. 10).

Because models of bonus bidding usually take positive bids as an assumption, they ignore an important element of bonus bidding that is not present in either profit-share or royalty bidding: the amount bidders will pay for the right to explore and develop the lease. A firm's market value depends on both the market value of the firm's tangible assets and the present value of opportunities for future investment. In oil and gas leasing, firms acquire much more than the potential to strike a "reserve" of oil. They also acquire valuable options to undertake discretionary investment. These are options the firm may or may not choose to exercise. If the leasing subject is bonus bidding, then the firm must anticipate both the NPV of the oil and gas in place and the value of the option to acquire the growth opportunities associated with the lease. Competitive bonus bidding requires the lessee to compensate the lessor for this option to explore and develop the lease. When a lease is awarded to the highest profit-share bidder, however, the lessee obtains a no-cost option to explore and develop the lease. This results in a decrease in the potential economic rent captured by the lessor.

Orthodox models of competitive bidding under uncertainty assume that an object with an unknown but unique value is auctioned off to a known number of bidders. In these models, it is assumed that bidders have independently and identically distributed unbiased estimates of the unique value (NPV) of the lease. Furthermore, some models assume that the competing firms are identical. That is, the auction consists of a number of firms with identical costs, identical attitudes toward risk, and identical expectations with respect to the value of the lease. Under these circumstances, it is appropriate to conceptualize the lease as having a definite but unknown value. Consider, however, two firms differing with respect to their engagement in research to discover lower-cost exploration techniques. The firm anticipating greater future success in the development of more-efficient exploration techniques will bid more for the option to explore and develop an oil and gas lease. Clearly, there are many ways in which firms with different attributes can capitalize on the options to engage in discretionary future investment. For this reason, it is unrealistic to assume that a given lease has a unique value to many bidders.

A commonly cited drawback of pure bonus bidding is the large front-end payment borne entirely by the lessee. It is often said that if capital markets are imperfect, then small firms will be at a disadvantage in raising the funds for the bonus-bid payment. However, firms will face different borrowing constraints even if capital markets are perfect. In the practical world of commerce, a common rule of thumb for evaluating debt policy is to consider the ratio of debt to the book value of equity. This appears to conflict with the prescription of modern finance telling us to consider ratios based on market values rather than accounting values. The redeeming element in applying the rule of thumb, though, is that book values refer to the value of the firm's tangible assets. But much of the value of firms engaged in exploration is accounted for by options to make future investments. "In general, tangible assets will be financed with more debt than investment options."[2] The presence of debt in the firm's capital structure attenuates the incentive to accept projects with positive NPVs. If the debt matures after the initiation of exploration, then bondholders will share the return on the investment with the shareholders.

Economic Analysis of Bonus-Bidding Results

We have analyzed the record of bidding for and production from 1,223 oil and gas leases issued by the federal government over the years 1954 through 1969 in the Gulf of Mexico area of the OCS. These leases are not pure bonus bids. Rather they all required payment of one-sixth royalty as a marginal cost. Consequently government policy failed to maximize the available economic rent. In terms of figure 3–1, the height of the bar representing the available total revenue is less than it would have been with pure bonus bidding.

1. Is bidding for OCS leases effectively competitive? In order to answer this question, we have computed the internal rate of return (IRR) earned by lessees on their investments in oil and gas leases. Our analysis is based on exact data for bonus, rental and royalty payments to the government, known gas and liquids production through the year 1979, and known drilling records. We have estimated exploration, drilling, and production costs. Further, we

have forecast production and revenue together with operating cost through the economic closedown point, not later than the year 2010.

If competition is inadequate, we would expect to find evidence in a rate of return that exceeds normal returns to equity capital for investments of comparable risk. We find no evidence of returns in excess of competitive yields. The results of our analysis are shown in table 3–1. On an after-tax basis, lessees earned a 10.74 percent rate of return on their lease investments. This is clearly not in excess of the 11.8 percent return on equity capital investments in all manufacturing corporations over a comparative time period.

It is evident that investments in oil and gas leases are not less risky than the "all-manufacturing" investment class shown in table 3–1. Our findings indicate that 61.9 percent of the 1,223 leases were dry and therefore yielded losses. Another 16.3 percent were productive but unprofitable in the sense that their costs exceeded their revenues using a zero discount rate. Only 21.8 percent of the leases were profitable. Thus, risk in oil- and gas-lease investments appears to be at least as great as for our benchmark investments.

Some concern has been expressed that large firms, easily able to overcome the entry barrier associated with the front-end bonus payment, might enjoy a competitive advantage over smaller firms. Any such advantage would be reflected in a higher rate of return earned by large relative to small firms. The results do not support such a concern. Table 3–2 shows that the Big 8 firms

Table 3–1

Comparison of Rates of Return on Equity Investment in 1,223 Oil and Gas Leases and all U.S. Manufacturing Corporations

(percent)

	Oil and Gas Leases (IRR)	Average Rate of Return on Equity Capital, All Manufacturing Corporations, 1954–1980
Before-tax return	13.16	20.2
After-tax return	10.74	11.8[a]

Source: Federal Trade Commission, *Quarterly Financial Reports of Manufacturing Corporations,* various issues.

[a]The standard deviation is 2.1 percentage points, with values ranging from 8.6 percent in 1958 to 16.5 percent in 1979.

Note: Our IRR estimates differ from the average rate of return to equity in manufacturing in the way they are computed and the purposes for which they are used. We have chosen to compare these two magnitudes because the rate of return to equity in manufacturing industries is, in our judgment, the best time series available to indicate rate-of-return levels for equity investments in a general sense. Note that our IRR estimate is relevant for the equity portion of investments in OCS leases only, and does not reflect the return to total capital invested in the OCS.

Table 3-2
Economic Data for 1,223 OCS Oil and Gas Leases Issued in the Gulf of Mexico, 1954-1969

	Aggregate Internal Rate of Return (percent)		Average Present Value per Lease after Taxes[a] (dollars)		
	Before Tax	After Tax	10 Percent Discount Rate	12.5 Percent Discount Rate	15 Percent Discount Rate
All 1,223 leases	13.16	10.74	115,575	−192,128	−334,893
Big 8 firms	12.82	10.37	54,968	−221,873	−349,795
Big 9-20 firms	13.62	11.26	218,734	−150,770	−325,116
All other lessees	13.63	11.15	181,706	−145,763	−295,255
Solo bids	12.24	10.10	14,392	−228,436	−339,515
Joint bids	14.74	11.74	356,251	−105,761	−323,897

[a]Base year is 1954.

earned a rate of return (10.37 percent) below that of either the Big 9 to 20 firms (11.26 percent) or all other lessees (11.5 percent).[3]

Because the analysis requires an aggregation of leases into groups, there is no means of testing whether different IRRs are statistically significant. As a supplementary approach, we undertook a multiple regression analysis to determine whether the logarithm of the high bid by lease differs significantly between Big 8 firms and all others. The big-firm bidding-advantage hypothesis would argue that large firms obtain leases at lower cost compared to smaller firms. Holding other factors in the regression constant, we find evidence of the opposite. The Big 8 firms paid significantly more than others for their leases (Mead, Mosiedjord, and Sorensen 1982).

Further, there has been some concern that joint bidding among firms competing for OCS leases might lessen competition. Table 3-2 also shows that firms bidding jointly earned a rate of return one percentage point above the average of all lessees; however, that return was still below the average yield on equity investments in all manufacturing industries.

The supplementary regression analysis tested the hypothesis that jointly bidding firms had a competitive advantage, with the result that prices paid by them were lower than for firms bidding solo. Holding other factors constant, we found no significant relationship between joint bidding and the price paid for oil and gas leases.

In sum, the evidence available indicates that bonus bidding, as used by the federal government, is effectively competitive in spite of the fact that the up-front money requirement may be a barrier to entry for small firms. It is not obvious that firms so small that the up-front bonus requirement would preclude entry are appropriate participants in OCS drilling and production. Oil and gas operation in the marine environment requires a high degree of technical competence and financial responsibility. From an environmental perspective, one might argue that the public is entitled to expect prompt action to control any accidental oil spill, such as occurred offshore from Santa Barbara in 1969, to clean up the spilled material, and to compensate those who are damaged. Firms too small to meet the bonus payments may not be in a position to satisfy this public interest.

2. Does the federal government collect the available economic rent under bonus bidding? Table 3-2 also presents data on the present value of several classes of leases under discount rates ranging from 10 to 15 percent. Using a 12.5 percent discount rate where all costs and projected future revenues are expressed in nominal terms, the analysis shows that the federal government collected more than 100 percent of the available rent.

3. Does actual production from leases correspond with bid prices? In a more-general context, this question asks about the rationality of bidding and performance under the bonus-bidding system. Results shown in table 3-3 indicate that the percentage of dry leases declines as the average bonus bid increases. Further, the average value of production per lease increases with higher prices bid for leases. These results indicate a basic rationality in bonus bidding in spite of the fact that there is a lower degree of correlation between the economic rent payment and the value of a lease under bonus bidding than under either profit-share or royalty bidding.

Due to the necessary aggregation of data in computing rates of return, we have no means of testing for significance of differences in average values shown in table 3-3. The regression analysis established that, holding other

Table 3-3
Relationship of Bid Prices to Bidding and Production Results

Bonus Bid Class	Number of Leases Issued	Percent Dry Leases	Average Bonus per Lease (dollars)	Undiscounted Average Gross Value of Production per Lease, Actual Through 1979 (dollars)
$250,000 or less	354.00	81.36	126,450	4,996,521
$250,001–$1,000,000	367.00	64.58	524,998	9,660,978
$1,000,001–$3,250,000	285.00	48.77	1,874,621	20,807,215
More than $3,250,001	217.00	41.01	9,002,511	42,295,839

factors constant, a significant positive relationship exists between the logarithm of the present value of oil and gas production through 1979 and the log of the high bid for leases. This evidence further confirms the basic rationality of bonus bidding.

Profit-Share Bidding

Administrative Costs of Profit-Share Leasing

Under profit-share bidding, each bidder offers a percentage of the net profit to be paid to the lessor for each lease. There are four possible advantages of profit-share bidding.

First, it is an improvement over royalty bidding in that the profit share to be paid is based on net rather than gross income. This means that as a field approaches exhaustion, its profit declines toward zero. Unless the profit-share bidding system also requires a fixed royalty payment, the problem of premature abandonment is avoided.

Second, profit-share bidding avoids the front-end loading problem that is characteristic of bonus bidding. No payments are due until production appears and profits accrue to the operator. In the absence of front-end payments, smaller, less well-financed operators may enter the bidding competition and possibly win leases.

Third, payments correspond with benefits. Dry holes require no profit-share payment. Conversely, the occasional rich discovery producing high profits results in larger payments to the government. This would avoid some of the political embarrassment associated with the Prudhoe Bay situation where a relatively small bonus payment was paired with an extremely rich discovery.

Fourth, a pure profit-share bidding arrangement may constrain over-zealous regulators and environmentalists from imposing uneconomic costs on oil exploration and production. Under the bonus payment system, any economic waste is clearly shared by the operator and the government.

There are also substantial disadvantages in the profit-share bidding system. First, although computation of profit may seem to be a simple and straightforward calculation, in fact, a wide variety of accepted accounting procedures are used. Two general concepts of profit may be identified. (1) Bidding may take place on an accounting concept of profit in which fixed costs and overhead costs, along with all operating costs, are deducted from gross revenue in order to obtain net profit. (2) Net profit may be defined as operating profits in which fixed costs and overhead are not allowed as reportable expenses.

Each general classification has a multitude of problems. For example, (1) where the lessee is an integrated producer and operates refineries, an arm's-length sale of oil or gas does not take place. In this case a transfer price must be determined. This involves subjective judgment. (2) The process of trading oil between separate companies is widespread in this industry. Where trading takes place, transfer prices must again be determined. (3) Where a company wishes to do some R&D concerned with oil exploration and production, it is likely to do so on leases involving profit-share payments rather than on other company production. (4) Where a company has a mixture of high-quality and low-quality drilling rigs or drill ships, it is likely to use the poor equipment on the profit-share lease and reserve the best equipment for other company operations. (5) Where a company needs to train crews in drilling and reservoir development, it is likely to do its training on profit-share leases. (6) "Gold-plating", where the share paid to government is high and the retained share is low, is likely to occur on profit-share leases. Evidence of this practice may be found in the Long Beach (Wilmington) field where profit shares paid to the government are extremely high. (7) In profit-share leases, public-relations expenditures are likely to be high. This will be particularly true when expenditures for public relations produce benefits for the lessee company as a whole. (8) In the event of supply shortages such as occurred in the 1973 and 1974 Arab embargo, operators are likely to allocate available supplies to their nonprofit share leases first.

Companies differ in their level of efficiency. In order for the lessor to select the highest bidder, it should evaluate probable efficiency for each competing bidder. This is a desirable practice, but it is also expensive and probably not feasible. But it means that the high profit-share bidder is not necessarily the operator that will produce the most economic rent for the lessor.

In order to avoid problems, the lessor will probably determine that it must carefully police lessee operations on profit-share leases, an operation that requires additional administrative costs. In terms of figure 3-1, one should expect not only the necessary cost of production to be incurred, but also unnecessary costs as well. But these costs reduce the available economic rents. Also, the added policing function by the lessor involves additional administrative costs, which dissipate some of its economic rents. Further, because the interpretation of profit is difficult, one must expect litigation of disputes. This requires expensive attorney fees and court costs for both the operator and the government, expenses that further reduce or dissipate the available economic rents.

Second, in addition to the profit-share bid, one must also consider the corporate income tax, for it too is a profit-share payment. With percentage depletion allowance totally phased out for all integrated oil companies and reduced for smaller nonintegrated firms, the corporate income tax will approach a 46 percent effective rate at the federal level. Any state income taxes will increase this rate even further. Using a 46 percent corporate income tax paired with a 30 percent profit-share bid results in an effective profit-share (or tax) rate of 62.2 percent. Thus, out of every additional dollar saved through efficiency, the company retains 37.8 cents. Where an 80 percent profit-share bid is paired with a 46 percent corporate income tax, the effective tax rate is 89.2 cents. This leaves only 10.8 cents on the dollar as a reward for efficiency, an incentive too small to produce maximum efficiency. In terms of figure 3–1, expenses will be higher than necessary, with the result that economic rents available to the government are sacrificed.

Third, as in the case of royalty payments, profit-share payments discourage investments in intensive field management, including well workovers, pressure-maintenance projects, and secondary recovery investments. Some supermarginal investments will become submarginal and will be passed over. The lost economic rents are borne by all citizens in the form of resource waste.

Fourth, further experiment with profit-share bidding is not needed. In the Long Beach case, the profit-share bid for the largest operating interest (80 percent) amounted to 94.56 percent of accounting profit being paid to the lessor, which leaves 4.44 percent of the profit for the operator as an efficiency incentive. The operator, however, also receives a 3 percent operating fee, and the fee is calculated as 3 percent of total expenditures. Thus, the higher the level of expenditures, the higher the operator's fee. This leaves a net incentive for efficiency of only 1.44 percent before taxes. Given a 46 percent corporate income tax, the after-tax incentive for efficiency is only 0.78 percent. In effect, there is no efficiency incentive.

As a substitute, the Department of Oil Properties of the City of Long Beach has developed a permanent staff to supervise and police the operators. Administrative interference with the operation of the field becomes a necessity. The Long Beach-Wilmington contract provides that

> the City Manager . . . shall exercise supervision and control of all day-to-day unit operations . . . and . . . shall make determinations and grant approvals in writing as he may deem appropriate for the supervision and direction of day-to-day operations of the Field Contractor, and the Field Contractor shall be bound by and shall perform in accordance with such determination.

A spokesman for one operator has stated that "hassle after hassle has developed regarding charges to the net profits accounts" (Mead 1969, p. 121). All of the problems outlined above can be verified in the Long Beach situation.

In summary, while profit-share bidding avoids some of the problems present in both bonus bidding and royalty bidding, the administrative costs associated with it are high. Economic analysis clearly indicates that economic rents received by the public would be substantially lower under profit-share bidding than under bonus bidding.

Profit-Share Bidding with Agency Costs and Asymmetric Information

Under profit-share bidding, potential lessees bid away a share of the project's net income in exchange for the right to explore and develop the tract and to retain a profit share if the bid is less than 100 percent. If both parties to the profit-share contract are utility maximizers, then the lessee will serve the lessor's best interest only when it is in his best interest to do so.

The lessor limits opportunism on the part of the lessee by defining appropriate incentives for the lessee. In the process, the lessor will incur monitoring costs associated with limiting the lessee's opportunistic activities. In a competitive market for profit-share leases, it will be in the lessee's best interest to devote resources to assure the lessor that the lessee will abstain from such opportunistic behavior. In the jargon of finance, the latter costs are termed bonding costs. The agency costs of the profit-share contract are equal to the sum of the monitoring costs incurred by the lessor, the bonding costs incurred by the lessee, and the residual loss to the lessor associated with the divergence between the actions taken by the lessee and those that would have served the lessor's best interest.[4]

The market's ability to achieve a competitive equilibrium will be severely restricted if potential lessees are not separable into well-defined risk classes. In general, lessees will not disclose their true attributes to the lessor. For this reason, profit-share bidding will be characterized by asymmetric information. Unless profit-share bidding is modified by complex contractual provisions, lessees will have a strong incentive to misrepresent their true risk characteristics. There is an extensive literature in economics and finance that attempts to isolate the necessary and sufficient conditions required to resolve the market failures associated with the presence of asymmetric information.[5]

Because profit-share contracts necessarily engender agency costs and asymmetric information, it is inappropriate to assume that different lessees bidding the same profit-share bid will produce equal absolute profit shares for the lessor. It follows that the lessor's claim under a profit-share leasing system will be contingent on the attributes of the lessee.

These costs are not borne by the lessor under a pure bonus-bid leasing system. In the case of competitive bonus bidding, the highest bidder wins the lease. The lessor receives the high bid as a guaranteed fixed contractual payment in exchange for the property rights to the lease's uncertain income stream.

For these reasons, we do not assume that the NPV of the lease will be the same under bonus and profit-share leasing systems. On a lease-by-lease basis, the lessor does not share the project's losses under a profit-share leasing system; however, there is no way to guarantee this when agency costs and asymmetric information are present. Although profit-share bidding constitutes a risk-sharing contract when profit is positive, the magnitude of the profit-share bid alone conveys no information about the lessee's potential for postcontractual opportunism. In fact, it is conceivable that the winning profit-share bid could equal or exceed 100 percent.[6] Under profit-share bidding with competing lessees having varied efficiency attributes, there is no economic theory of how operating efficiency corresponds monotonically with the percentage profit share bid. Orthodox models of lessor rent capture have typically either ignored or minimized the impact of agency costs and asymmetric information on the efficacy of profit share leasing.

Leland (1978) does discuss asymmetric information, the problems of defining an accurate profit base, and the administrative costs of monitoring the lessee under profit-share leasing. His model, however, does not deal with the problem of discriminating between lessees of differing efficiency. If profit-share leasing is to be efficient, then the problems of agency costs and asymmetric information must be addressed *prior* to the award of the lease. This will require the lessor to impose side constraints on each bidder, which will, in general, be different for lessees with different efficiency attributes. Thus, even under strong assumptions about the accounting information available to the lessor with respect to potential lessees' efficiency attributes, profit-share bidding cannot be conducted in a competitive manner. Each bidder will submit a profit-share bid subject to a different set of side constraints. The winning bidder will not necessarily be the high profit-share bidder.

McDonald (1979) also considers the problem of auditing the lessee's financial records, but he does not consider how the lessor should *choose* a lessee under profit-share bidding, even if accounting records are perfect.

Ramsey (1980) notes the problems associated with defining costs such as the return-to-equity capital, as well as the problem of allocating overhead costs if the lessee has more than one lease. He argues that the objection of profit-share leasing on the basis of its incentive effects on cost minimization at high profit-share percentage rates is, under competitive circumstances, illegitimate. Note, however, that profit-share leasing can be competitive only prior to the award of the contract. After the lease is granted, there does not exist free entry into the exploration and development phases of the venture. It is meaningless, therefore, to refer to a "competitive rate of return" after the lessee is awarded the right to operate the lease.

Reece (1978, 1979) assumes that the lessor does not share any losses of a profit-share lease. In the long run, however, the lessor pays for dry-hole costs because these must be paid out of productive leases. He also assumes that the tract's true gross-of-bid value is independent of the leasing system used. These assumptions imply that society will capture the same amount of total rent regardless of the leasing system implemented. Clearly these assumptions ignore all of the problems of agency costs and asymmetric information. Therefore, it is impossible to determine how the lessor would award the lease under a more-realistic scenario in which firms have differing costs, risk preferences, and expectations. Reece (1979, p. 48) concludes that government rent capture will be substantially greater under a profit-share bidding system than under a bonus-bidding system. To achieve this result, however, it was necessary to invoke assumptions that served to ignore the efficiency issue altogether.

Sebenius and Stan (1981) derive a risk ranking for alternative leasing systems, while assuming that the incentive problems of profit-share leasing have been solved. They do not discuss the impact of agency costs and asymmetric information on the lessor's revenue collections.

Models of profit-share leasing are severely handicapped in their ability to incorporate the impact of entrepreneurial anticipations. Under bonus bidding for the right to explore and develop a tract, entrepreneurs have an incentive to capitalize on their investment in technological R&D by purchasing options on oil and gas tracts in the present. How is this accomplished in the perfect-information models of profit-share leasing? Clearly the profit-share bid will influence the lessee's incentive to invest in the development of more-efficient methods of exploration and development.

Agency and asymmetry problems do not exist in the pure bonus-bid leasing case because the lessor has accepted a fixed contractual payment in exchange for all of the property rights to the lease's uncertain income stream. The lessor's income is fixed regardless of the slings and arrows of fortune. Hence, if bonus bidding for the right to explore and develop the tract is competitive, then the winning bonus bid will be a sufficient statistic of the lessee's efficiency attributes. Because all of this complex information about risk preferences, costs, and expectations is summarized by a scalar variable (the bonus bid), it is a simple matter for the lessor to determine the optimal contract. If the lessor's objective is to maximize economic rent collection, then the optimal course of action is to accept the highest bonus bid. If the lessor has an objective other than the maximization of economic rent, then by comparing the rent under the alternative scheme with the potential rent of the high bonus bid, it is possible to evaluate the economic cost of alternative leasing arrangements.

In the profit-share leasing case, the optimal contract will consist of a vector of components. Such a contract must specify not only the profit share, the time to the contract's maturity, and the dimensions of the lease but a host of other highly ambiguous variables as well. These include the definition of all costs; the specification of auditing procedures; the required work commitment; constraints on the time profile of lease development; the lessee's decision-making authority; and conditions under which the contract may be prematurely terminated by either party. Thus, although it might be possible to specify the elements of an optimal profit-share bid contract under certain ideal conditions, it will remain unlikely that one bid offer will be superior to all others in every component. Therefore, unless we are willing to specify a utility function for the lessor, it will generally be impossible to rank the alternative competing contracts.

Under a competitive bonus-bid leasing system, agents are disciplined by the vicissitudes of competition to bid away all of the anticipated rents of the lease. Our empirical data indicate that under the bonus-bidding system with a fixed 16 2/3 percent royalty, the federal government has collected approximately 100 percent of the available economic rent. Of course, the winning firm under profit-share bidding is not necessarily the firm anticipating the largest NPV for the lease.

Some Evidence from the Long Beach-Wilmington Oil-Field Experience

This section provides a sketch of findings drawn from interviews with senior executives at the THUMS Long Beach Company, the Long Beach Department of Oil Properties, the State Lands Commission (Long Beach, California), the Long Beach Oil Development Company, the Western Oil and Gas Association, Chevron Oil Co., Union Oil Co., and Mobil Oil Co.[7] We also refer to findings drawn from a review of the internal controls and operations of the THUMS Long Beach Company conducted by Deloitte, Haskins, and Sells (DHS)(1981). The review was completed under a contract from the State Lands Commission (California).

Under the leasing arrangement between the state, the city of Long Beach, and the THUMS Long Beach Company, the city is assigned responsibility for the supervision of day-to-day operations, accounting, and the payment of oil and gas revenues to the state. The THUMS Long Beach Company was created by Texaco, Inc., Humble Oil Co. (Exxon), Union Oil Co., Mobil Oil Co., and Shell Oil Co. It is a nonprofit corporation and is responsible for all exploration and development activities. It is, however, subject to the oversight of the city of Long Beach Department of Oil Properties, as well as the State Lands Commission. The Long Beach Unit is a large-scale oil-field operation. Production from more than 600 wells yields over $600 million in annual revenues (1980).

The Long Beach Operations Group of the State Lands Commission is responsible for the Long Beach Unit's extractive development. To accomplish this, the Long Beach Operations Group is divided into the following departments: economic research, capital projects, geology, petroleum engineering, and reservoir engineering. The state votes on issues coming before the voting parties and is also responsible for approving the annual plan of development and operations.

There is a consensus among city, state, and industry officials that the Wilmington Field comprised an extremely well-defined field prior to profit-share bidding. The field had a "known area, a known volume, and core holes had been drilled prior to bidding." One industry official said that "reserves were known with 90 percent confidence." In fact, the field had been described as "a tank of oil." There was certainly "no doubt that the Wilmington Field would prove to be productive." With this in mind, recall that a principal argument for profit-share bidding is that it enables the lessor to capture more of the economic rent when bidders are inclined to underestimate the lease's potential. When bidders strongly anticipate striking productive reserves, profit-share bidding does not serve to maximize the lessor's economic rent.

A second principal argument for profit-share bidding is that it permits the lessor to profit from large, unanticipated price increases. In light of the dramatic price increases of the 1970s, the city and state have certainly profited from this feature of profit-share bidding. Note, however, that the state and the city would have retained the option to impose additional taxes on the lessor if bonus bidding had been used. Thus, profit-share bidding is not a necessary condition for the lessor to share in large, unanticipated price increases. Furthermore, under profit-share bidding the lessor shares the risk of price decreases. In fact, the State Lands Commission devotes considerable resources to forecasting crude prices. A $2 per barrel decline in price translates into a $40 million decline in revenue for the city and the state when annual production is at 20 million barrels per year. When crude prices are volatile, it becomes impossible for the city and the state to form accurate forecasts of annual revenues. Thus, when prices drop unexpectedly, there exist incentives for the state, as a voting party, to accelerate the rate of development in order to maintain the anticipated level of revenues.

General Management

According to the DHS report, there is no formal statement defining long-term objectives for the Long Beach Unit. Nor is there a statement outlining procedures for resolving conflicts between the city, the state, and the field contractor. For this reason, there is excessive interference in the field contractor's day-to-day operating decisions. Industry officials dispute this finding. They point out that Long Beach has ultimate responsibility for the direction of day-to-day operations as the unit operator. They also cited the numerous agreements that exist to ease the working relationship between the city and THUMS. Several industry officials said that the real source of day-to-day interference in the field contractor's operating decisions is the state's control of operations through the annual budget. At times the state has controlled field operations by withholding funds from the budget, using an augmentation and modification process to dole out money a little at a time.

Management Control

According to the DHS report, management control reports show historical information and not progress toward operational goals. There are no performance measures of real operating results. In particular, performance measures of drilling operations are not maintained. For example, there are no performance measures of time per foot drilled, rig days per well, or rig downtime per accounting period. According to DHS, there do exist operating reports for monitoring short-term operations; however, these are not summarized so as to make them useful in gauging performance for longer periods.

Reservoir Engineering

DHS report that although it is standard industry practice to evaluate the economic life of the field as well as the optimum rate of lease development by conducting an annual reservoir study, no such evaluation is made for the Wilmington unit. Furthermore, DHS report that plans for wells, cellars, rigs, facilities, and personnel are not coordinated. Industry officials claim that these findings are without merit. They cite numerous reports to the city and the state outlining long-term development plans.

Development Engineering

DHS report that "the drilling approval process is lengthy and inefficient by industry standards." They also found that the payback method rather than the IRR or the NPV is used to evaluate investments.[8] Industry officials argue that "it might not be economical" to reduce the drilling approval process to less than twenty weeks. One official pointed out that when the payout period on some investments is as short as two months, there is no need to use the NPV.

Production Engineering

Independent production engineering staff are located at the city, the state, and the field contractor. DHS cite this as evidence of duplication in engineering efforts. An industry official acknowledged some duplication here but noted that it amounted to only $143,000 per year for a $600 million per year operation.

Production Operations

DHS report that the job-cost system used for direct well charges excludes material and overhead costs. For this reason, it is impossible to determine accurate total costs for well servicing jobs.

The purpose of this discussion is not to attempt to summarize the DHS findings. In fact, we have cited only a small fraction of the body of DHS findings. Our objective is to show that such a review can impose significant administrative costs on the lessee and the lessor. These are social costs for the nation. Currently the city, the State Lands Commission, and the THUMS Long Beach Company are performing independent reviews of the DHS recommendations. Clearly the lessee incurs significant bonding costs to assure the lessor that the lease is managed efficiently. Similarly the lessor incurs significant monitoring costs. Because many of the day-to-day operations of the city and THUMS are dependent on verbal agreements, it is particularly difficult to perform a competent review of the lessee's operations. Furthermore, for such a review to be of value, it must be based on a norm. Any accounting norm, however well conceived, will be based on historical information, whereas the economic norm of survival in competitive markets with free entry requires superior anticipations of future market conditions.

To prevent the unrestrained transfer of experienced personnel by the major oil companies from the Wilmington Field, the THUMS Long Beach Company has been set up as an autonomous entity. There is no transfer of management from THUMS to the major oil companies. The management of THUMS are career personnel. The formation of such an independent company for a net profit-share lease clearly serves to reduce some of the participant's incentives to engage in opportunistic behavior; however, the operating company generally will bear little or no relationship to the bidding companies. Hence, the formation of such a company defeats the purpose of the bidding system: to choose the efficient lessee. Further, when as many as five major firms acquire working interests in the lease, each firm may have a different operating philosophy. Thus, in the words of one official, "The operating company will require a strong manager," and it is not at all clear how such a manager should be chosen from among the competing profit-share bidders.

To summarize, there is considerable evidence of substantial administrative costs associated with profit-share leasing in the Long Beach-Wilmington oil field, which would be unnecessary in a bonus-bid situation. The state and the city incur considerable costs of monitoring the THUMS Long Beach Company, and THUMS in turn incurs costs associated with assuring efficient operation of the lease. The state and the city could have profited from unanticipated price increases under a bonus-bid leasing system by raising ad valorem taxes. Under profit-share bidding, the city and the state share the risk of a decline in the price of crude. For this reason, the state devotes resources to forecasting crude prices in order to anticipate annual revenues. Clearly if profit-share leasing were to be implemented on a large scale on the OCS, then the rate of lease development could easily become the subject of political bargaining. No longer would the objective of leasing policy be to maximize and collect the present value of economic rent. Industry officials agree that the implementation of profit-share leasing will require the formation of a separate company to operate each lease. However, if the operating company is truly severed from its parent(s), then it is difficult to say just what bidding for a profit-share lease has accomplished. The efficiency of the operating company need not bear any relationship to the parent company(ies). The Long Beach-Wilmington Field was generally expected to be a highly productive lease. There is no evidence to indicate that the lessor would not have collected all of the economic rent under a bonus-bid leasing system. The problems of accumulating the quantitative data to evaluate operating efficiency, ex post, appear to be insurmountable. There does not exist a secondary market for operating shares in the lease. In the case of OCS leases acquired by bonus bidding, lessees have the right to sell the option to explore and develop the lease. The existence of such a secondary market facilitates the entry of efficient lessees when the original lessee has miscalculated the lease's NPV. In the case of profit-share leasing, lessees are not penalized for overestimating their operating efficiencies.

Conclusions

If mineral-leasing policy is to serve the public interest, the primary objective of the government should be to maximize and then collect the full present value of the economic rent available from its resources. We have evaluated the cash bonus and the profit-share bidding alternatives from both theoretical and empirical perspectives.

From a theory perspective, we find a pure cash-bonus-bidding system to be superior on balance to profit-share bidding. The overriding advantage of bonus bidding is that it harmonizes private incentives to minimize costs relative to any revenue stream, with the social-welfare goal of maximizing the economic rent. In contrast, due to what we have termed agency and information asymmetry problems, operations under a profit-share regime lead to losses in economic rents as lessees pursue their own interests, which vary from the public interest. Even selection of the lessee from among the competing profit-share bidders is faulty. Firms differ in levels of efficiency, and the government cannot be sure that the firm bidding the highest profit share will operate efficiently and thereby create the largest absolute level of profit to be shared.

From an empirical perspective, we find that oil and gas leases issued by the government between 1954 and 1969 have been effectively competitive and that the government collected at least 100 percent of the available economic rent. In practice, the bonus-bidding system used by the federal government is not a pure bonus system but requires a one-sixth royalty payment in addition. This feature reduces the available economic rent because it leads to premature abandonment of leases and reduces socially desirable investment in greater amounts of oil or gas recovered.

Profit-share leasing leads to excessive administrative costs for lessees, thereby reducing the available economic rent. Further, lease-sale administration by the government in the Long Beach-Wilmington oil field is shown to be inefficient, thereby dissipating some of the economic rents. Under profit-share leasing, the state and the city bear the risk of crude-oil price decreases as well as cost increases. This is not the case for a bonus-bid leasing system. The state and the city could have profited from unanticipated price increases under a bonus-bid leasing system by raising ad valorem taxes. The lessee would have borne the losses associated with unanticipated price decreases. The implementation of profit-share leasing requires establishment of an autonomous company to operate each lease. If the severance of the operating company from the parent company(ies) is complete, then the efficiency attributes of the operating company need not bear any relationship to those of the parent company(ies).

There is no evidence to indicate that the state and the city would not have collected all of the economic rent under a bonus-bid leasing system. The Long Beach-Wilmington Field was generally expected to be highly productive. The implementation of profit-share leasing engenders agency and asymmetric information costs, as well as administrative costs. The problems of accumulating the quantitative data to evaluate operating efficiency are enormous. We have shown that profit-share leasing entitles the winning bidder to a no-cost option to explore and develop the lease. OCS leases acquired by bonus bidding require the lessee to pay for this valuable right. If the lessee has miscalculated the NPV of the lease, then the existence of a secondary market for OCS leases facilitates the entry of efficient lessees. In the case of profit-share leasing, such secondary markets for the right to explore and develop the lease do not exist. Thus profit-share bidders do not bear the social costs of overestimating or misrepresenting their operating efficiencies.

Notes

1. For a more complete development of this topic, see Pickett (1982).

2. See Myers (1977) for a theoretical development of this important result.

3. Defined in terms of worldwide sales revenue.

4. See Jensen and Meckling (1976) for an excellent discussion of agency costs and managerial behavior. Haugen and Senbet (1979, 1981) show how agency problems might be resolved by option contracts.

5. If all projects are initially valued at some average amount, then the presence of asymmetric information ultimately leads to the failure of venture-capital markets (Akerlof 1970). Ross (1977) examines ways in which the firm can be induced to reveal accurate information about its risk class and therefore its associated return stream.

6. The Long Beach contract allows an operator's fee. Part of this might be bid away. An operator with nearby refineries with long-term idle capacity might be willing to bid more than 100 percent in order to obtain committed oil and minimize refinery losses. Further, an operator might believe that oil could be charged out of the lease (into the refinery) at below the opportunity cost of oil. The revenue loss would be borne by the lessor, and offsetting gains would accrue to the lessee as the buyer of the oil. These are examples of agency costs in a profit-share bidding framework.

7. The interviews were conducted either by telephone or in person. Permission for the interviews was granted with the understanding that no person would be quoted directly. All of the officials interviewed are either involved in litigation or apprehensive about disclosing information that could lead to costly litigation. For this reason, it was impossible to acquire the kind of quantitative data on operating costs that might permit an economic comparison of operating efficiency in the Long Beach-Wilmington Field with that of the OCS. The Long Beach Oil Development Company is not involved in the administration of the Long Beach-Wilmington Oil Field.

8. The payback method does not account for the opportunity cost of money.

References

Akerlof, G.A. 1970. "The Market for 'Lemons': Quality Uncertainty and the Market Mechanism." *Quarterly Journal of Economics* 84 (August): 488–500.

California. Office of the Auditor General. 1977. *Financial Audit of the Tidelands and Submerged Lands Held in Trust by the City of Long Beach for the State of California Year Ended June 30, 1976.* California Legislature Report 708.

Deloitte, Haskins, and Sells. 1981. *Review of the Internal Controls and Operations of the THUMS Long Beach Company: Final Report.* State Lands Commission. April.

Haugen, R.A., and Senbet, L.W. 1979. "New Perspectives on Informational Asymmetry and Agency Relationships." *Journal of Financial and Quantitative Analysis* 14 (November): 671–694.

———. 1981. "Resolving the Agency Problems of External Capital through Options," *Journal of Finance* 36 (June): 629–647.

Jensen, M.C., and Meckling, W.H. 1976. "Theory of the Firm: Managerial Behavior, Agency Costs and Ownership Structure." *Journal of Financial Economics* 3:305–360.

Leland, H.E. 1978. "Optimal Risk Sharing and the Leasing of Natural Resources, with Application to Oil and Gas Leasing on the OCS," *Quarterly Journal of Economics* 92 (August): 413–437.

McDonald, Stephen L. 1979. *The Leasing of Federal Lands for Fossil Fuels Production.* Baltimore: Johns Hopkins University Press.

Mead, Walter J.; Moseidjord, Asbjorn; and Sorensen, Philip E. 1982. "Competition in OCS Oil and Gas Lease Auctions—A Statistical Analysis of Winning Bids." Forthcoming.

Mead, Walter J. 1969. "Federal Public Lands Leasing Policies." *Quarterly of Colorado School of Mines* 64 (October): 212.

Myers, Stewart C. 1977. "Determinants of Corporate Borrowing." *Journal of Financial Economics* 5 (November): 147–175.

Pickett, Gregory G. 1982. *An Option Pricing Model of Bonus Bidding and Profit Share Bidding for Offshore Oil and Gas Leases.* Unpublished Ph.D. Dissertation (University of California, Santa Barbara, Calif.).

Ramsey, F.B. 1980. *Bidding and Oil Leases.* Contemporary Studies in Economic and Financial Analysis, Vol. 25. Greenwich, Conn.: JAI Press.

Reece, D.K. 1978. "Competitive Bidding for Offshore Petroleum Leases." *Bell Journal of Economics* 9 (Autumn): 369–384.

———. 1979. *Leasing Offshore Oil: An Analysis of Alternative Information and Bidding System.* New York: Garland Publishing.

Ross, S.A. 1977. "The Determination of Financial Structure: The Incentive-Signalling Approach." *Bell Journal of Economics* 10 (Spring): 23–40.

Sebenius, J.K. and Stan, P.J.E. 1981. "Risk-Spreading Properties of Common Tax and Contract Instruments." *Rand P-6600* (March): 1–26.

Smith, C.W., Jr. 1976. "Option Pricing: A Review." *Journal of Financial Economics* 2: 3–51.

4

Encouraging Solar Water Heating: Some Implementation Issues

Karl Hausker and
Eugene Bardach

Since 1979, the California Public Utilities Commission (PUC) has conducted an ambitious program to accelerate growth in the demand for solar domestic hot-water heating (swh). In doing so the utility regulators have undertaken tasks far different from those traditionally entailed in regulating utility rates and performance in a quasi-judicial manner. This chapter describes and evaluates the way in which the PUC has implemented this essentially promotional program.

A second purpose is to evaluate several alternative strategies governments can use to stimulate the swh market. California has experimented with many such strategies in addition to the PUC program—for example, solar tax credits, local solar mandates for new construction, municipal solar utilities, and miscellaneous informational and promotional activities. In addition to the PUC program, in this chapter we focus on the state solar tax credit and local solar ordinances.

The Case for Promoting Solar Energy

The rationale for government intervention in the solar market begins with the fact that the marginal cost of energy supplied by a utility is generally greater than the average cost of all utility supplies. In water-heating applications, utility marginal costs can be higher than the cost of solar energy. If private parties faced the marginal cost of utility services, they would sometimes find it more economical to invest in swh. However, rate-of-return regulation of utilities rolls in the cost of expensive marginal sources with the lower cost of other sources. The consumer pays only the average cost, and compared to this cost swh does not appear to be economical.[1] Government intervention to promote these more-efficient investments, therefore, can potentially furnish net social benefits.

Most energy analysts agree that solar energy makes sense when displacing electric-resistance water heating. Several reputable studies show swh to be less costly than electric water heating for society as a whole. The recent study "Solar Energy in the 1980s" estimated the life-cycle societal cost of electric water heating to be about $8,800 in existing dwellings.[2] The life-cycle societal cost of swh with electric backup was estimated at $7,500 to $8,000. In contrast, the consumer would pay only $4,600 to $5,700 for electric water heating, and roughly $7,000 for solar water heating in the absence of any solar subsidies.[3]

Gas water heating presents a far different picture. Solar energy is uneconomical compared to gas from both the private and the societal perspective and for both new and existing dwellings. Gas water heating costs about $3,500 on a life-cycle basis from a societal perspective.[4]

Ideally government should encourage the use of swh only where it substitutes for electricity. Such encouragement could take many forms. State or local building standards could require the installation of solar water heating in new dwellings or upon the resale of existing dwellings. Municipal solar utilities could possibly lease equipment to home owners at a price lower than that faced by an individual buyer. Finally, government could deliver direct subsidies to swh buyers through tax credits, utility rebates, or subsidized loans. The PUC program, our focus here, is an example of this last policy design.

Although investments in swh can be cost-beneficial in many circumstances, they are not as cost-effective as investments in other sorts of energy-saving equipment. Ideally, society should not encourage investments in swh until it has pushed a number of other, much less-expensive, conservation measures a good deal further. Flow-restriction devices on hot-water outlets, for example, produce energy savings at less than one-third the cost of swh investments. Other measures that are considerably more cost-effective than swh are thicker water-heater insulation, increased clothes washer and dishwasher energy-efficiency, heat-pump boosters to electric water heaters, and conversions of electric water heaters to gas whenever feasible.[5] Nevertheless, government promotion of solar water heating is often politically more popular than most of these more cost-effective measures and in many circumstances may be administratively more feasible. Although we may hope that this situation will change, the move to support swh is a current reality. Many regulatory and subsidy programs are in place and are being considered. Therefore, we need to understand the issues that affect their design and implementation.

The Demonstration Solar Financing Program

In 1978, the California legislature instructed the California PUC to "investigate the feasibility of alternative methods of providing low-interest, long-term financing of solar energy systems for utility customers." This directive was consistent with the state's general commitment to renewable-energy development. It also reflected concerns over the allegedly slow rate of market penetration by solar energy systems and the perceived inequities of the existing solar income tax credit.

In January 1980, the PUC reported its findings to the legislature and through a number of subsequent decisions, launched a major program of utility-financed subsidies to solar water-heating systems, known as the Demonstration Solar Financing Program (DSFP). The program offered utility customers cash credits (rebates) or low-interest loans for the retrofit of solar water heating in existing residences. Ratepayers were to bear the cost of the program, and the four major investor-owned California utilities were to administer it. Table 4-1 summarizes the types and levels of subsidies ordered by the PUC, the market-penetration goals, and the estimated utility savings.

According to the PUC's estimates in table 4-1, the DSFP would have produced millions of dollars of savings to ratepayers had it been limited to installation where swh supplemented electric water heating (or electric retrofits). Instead of implementing such a program, however, the PUC became heavily involved in what one of us has elsewhere called a game of "Piling On," in which the "player" adds various objectives of dubious relevance and/or value to an underlying, legitimate policy concept.[6]

Piling On

By far the most harmful of the objectives piled on by the PUC was the goal of stimulating the retrofit of swh in homes using gas water heating (or gas retrofits). The DSFP aimed to spur gas retrofits in 40,000 single-family homes and in over 265,000 apartment units although gas retrofits are uneconomical from a societal perspective.

The PUC's own estimates showed that this is equally true from the utility's or ratepayer's perspective. The envisioned number of gas retrofits will produce an estimated net loss to ratepayers of $42 million.[7]

Another goal piled on by the PUC was the extension of benefits to low-income families. Concerned with extending the program to the less affluent, the PUC ordered the utilities to supply nearly 2,000 swh systems free to low-income families. The net cost was projected at $6.2 million. In this way, the PUC hoped to avoid a major criticism leveled at the solar tax credit: that it benefited mainly upper-income people. Two-thirds of those who claimed the tax credit in 1978 had gross adjusted incomes above $25,000.[8] Well over half the applicants for the credits used the solar installations to heat swimming pools.

The PUC program also included twenty-year loans at 6 percent interest to gas customers in two of the four utility service areas covered by the DSFP. The loan program had several apparent objectives. One was to assist low-income persons who might not have enough money to purchase a system with out-of-pocket cash and who also might not have the necessary credit rating to borrow from commercial lenders. A second objective was to buffer swings in

Table 4-1
Costs and Savings of the Demonstration Solar Financing Program

Target Market	Type of Subsidy	Subsidy per Dwelling	Market Penetration Target	Target of Percent of Total Market	Estimated Utility Savings ($000)[d]
Single-family, electric	Credit	$720[a]	70,940	15	48,050
Single-family, gas	Credit	$960[b]	21,000	0.5	−10,770
Single-family, gas	Low interest loan	6%, 20-year loan	18,500	0.5	−10,670
Multifamily, gas	Credit	$288 per apartment[c]	266,600	10	−20,310
Low-income families	Free solar system	100% of cost	1,780		− 6,260
Total			378,820		

Source: PUC Decision 92251, September 16, 1980, table III, p. 14c.

[a]Twenty dollars per month for thirty-six months. One utility's credit originally was structured differently.

[b]Twenty dollars per month for forty-eight months.

[c]Eight dollars per month per apartment unit for thirty-six months.

[d]This column is equivalent to the net utility revenue requirement column in Decision 92251. The figures indicate cost-effectiveness from the ratepayers' perspective and are equal to utility-avoided costs minus lost revenue minus subsidy payments. Administrative costs are not included. The PUC presented these figures in NPV terms at a real discount rate of 4.3 percent. For a full treatment of different perspectives on the cost-effectiveness of conservation and solar programs, see Kevin White, "The Economics of Conservation," *IEEE Transactions on Power Apparatus and Systems* PAS-100, no. 11 (November 1981): 4546–4552.

commercial lenders' policies toward making loans for swh systems. A third objective was to ascertain the extent of consumer interest in the two types of subsidies (loans and cash credits). The desire to reach low-income groups was also probably part of the commission's motivation in extending cash credits to the multifamily gas market (electric water heating in apartments is rare).

The Demonstration Rationale

Although on balance we are critical of the PUC's desire to extend the subsidy to gas customers and all income classes, the demonstration goals of the program could be thought to justify the extension. That is, if the goal was not merely to promote and stimulate solar but was to test the validity of some marketing strategy or tactic, then perhaps we should admit these demonstration goals to the definition of the basic policy goals and not stigmatize them as having been piled on. In our view, however, the DSFP was not actually much of demonstration.

First, there is circumstantial evidence, supported by interviews with the PUC staff, that calling the program a demonstration was in part a tactic intended to keep the PUC out of legal trouble. In 1979 the California State Supreme Court had prevented the PUC from ordering Southern California Gas to undertake a conservation financing program.[9]

Beyond the legal issue, the PUC genuinely believed that it was largely continuing its investigation into the feasibility of methods of utility-assisted financing for solar systems. The commission believed that creation of a vigorous, ambitious program was the way to answer the many questions it posed for itself:

> What is necessary to encourage consumer acceptance of solar water heating? Can the solar industry deliver adequate supplies at reasonable prices? Will the solar industry provide adequate quality and service? Will banks, savings and loans, and credit unions provide adequate and attractive financing? Should or must utilities play an active role in consumer protection? If so, what limits to utility activity are necessary to preserve competition? Without the demonstration programs, we would be reduced to sheer speculation on these and other questions of fundamental importance.

> We believe the best way to convince the public of the current viability of solar water heating is to stimulate the installation of a significant number of systems in neighborhoods throughout the state. When solar devices are in each neighborhood, it will become readily apparent that the technology is available now. When people hear by word of mouth about system performance, an industry reputation will be established.[10]

The implicit and explicit questions posed are certainly legitimate, and they are of interest to the PUC and the public at large. What is much less clear, however, is why a program like the DSFP was needed to answer them.

Consider first the demonstration to the public that the "technology is available now." Implicit here is the claim that it is available economically, yet this was true only for electric retrofits. Taking into account the PUC subsidy and the tax credit, the private costs of swh were slightly less than the private costs of electric water heating.[11] On the gas side, however, gas water heating proved to be over $1,000 cheaper than swh even for the homeowner receiving both subsidies.[12] As the PUC's report "Financing the Solar Transition" put it, "Based on the marginal cost of gas at the beginning of our investigation, the cost-effectiveness of retrofitting existing, gas water heaters with solar energy devices was dubious."[13] The report then went on to speculate, however, that natural-gas prices would rise sufficiently rapidly in the near future to justify the extension of the program to the gas market. If the object was to demonstrate that solar retrofitting was cost-effective for gas water heating, the PUC should have waited until it was sure that such a demonstration could actually be made in good faith.

As to the low-income population, if the objective was to demonstrate to poor people that persons like themselves could benefit from swh, the mere giving away of solar systems to 2,000 lucky families could hardly have accomplished the task. With regard to the choice between credits and low-interest loans, a much smaller demonstration program could have accomplished the goal of illuminating this choice. A subsidy program designed to reach several thousand home owners would have been adequate. In addition, limiting the loan program to gas customers made no sense at all in light of the fact that the prime target population, for which a constructive subsidy program might indeed have been developed, was the market of electric-water-heater owners.

Finally, we would acknowledge the significance of the many questions pertaining to the reliability and integrity of the solar industry. Some of these questions could have been answered simply by studying the industry as it had in fact evolved up to the point the DSFP was launched. The resolution of other questions, particularly those pertaining to consumer protection and its effects on the industry, depended on scaling up the industry in ways envisioned by the DSFP designers. The uncomfortable fact, however, is that there is actually no way to answer the questions about the effects of industry expansion to any specified scale without actually expanding the industry to that scale. To say that this is a demonstration is a misuse of the term.

The Consumer-Protection Quandary

Many of the implementation problems that plagued the program can be traced to the PUC's consumer-protection package. Rather than taking modest measures to give consumers some assurance of system quality, the PUC piled on the goal of molding the swh industry to fit the commission's conception of what a quality-conscious, reputable industry should be.

The PUC aspired to improve substantially on California Energy Commission standards developed as part of the state solar-tax-credit program. In order to qualify for DSFP subsidies, a swh installation had to meet a minimum-performance standard, meet certain prescriptive standards for system design and installation, and carry a five-year full warranty with partial coverage up to ten years. The PUC ordered utilities to conduct on-site inspections to assure compliance with these requirements. This consumer-protection package actually damaged the commission's own program goals and had a number of unintended effects.

The commission established the rigid performance standard of requiring all systems to displace at least 60 percent of conventional energy. This standard effectively excluded what may be the most cost-effective system design: a simple passive solar water heater supplying 30 to 60 percent of a household's hot-water energy needs. Although these systems produce less gross energy, their low cost and low maintenance requirements can often make them more economical than the more-common active swh systems. Some passive systems still qualify for the program, but only if several are hooked up in an effort to guarantee 60 percent displacement of conventional energy. This goal may prove elusive: passive systems are designed to produce less gross energy but at very low cost.

The prescriptive standards initially adopted by the PUC also tended to discourage installation of the most cost-effective systems. The best example of this effect was the commission's requirement that freeze protection be accomplished by means other than heat. The most common freeze-protection methods in California at the time did use heat (recirculation of hot water through collector pipes). Although this method may result in somewhat lower gross energy savings, overall system costs are lower than with other methods. In California's relatively warm climate, this is a rational economic trade-off. Once again, the commission was concerned more with gross energy savings than with cost-effectiveness.

PUC prescriptive standards were inflexible and detailed and would have required lengthy, expensive utility inspections. For example, the standards could conceivably have required utility inspectors to check solar-system compliance with fifteen different state and national codes ranging from electrical to plumbing to sanitation codes. Moreover, the standards were incompatible with many of the designs still competing in the market. Had they been enforced, they would have inhibited competition and stifled innovation.

Finally, the PUC initially proposed some of the most-demanding warranty requirements ever imposed on any industry: a five-year full warranty (repairs at no cost to the consumer) and an additional five-year prorated parts warranty. The PUC equated extended warranties with consumer protection without thinking through the question of how to deal with a young, emerging industry. It was surely unrealistic to impose warranty responsibilities that were to last a decade when the industry itself was less than a decade old. How could a firm rationally estimate its long-term, warranty-related costs and then reflect these in its price? Indeed the effects of unrealistic warranty requirements might have been contrary to PUC goals. Unscrupulous firms with no intention of staying in business might have garnered business by selling at a price that did not reflect any consideration of warranty responsibilities. Ultimately only the owner's concern for the reputation of a firm gives meaning to a warranty, be it voluntary or mandatory.

The potential direct effects of the PUC's original consumer-protection package were a loss of economic efficiency, a stifling of innovation, and the effects of unrealistic warranties. More-subtle adverse effects were also possible. The PUC made no serious attempt to integrate its consumer protection with existing state programs.[14]

In sum, the commission piled many questionable goals onto the justifiable policy objective of subsidizing electric retrofits. The PUC expanded the program to include gas retrofits, gave a veneer of equity to the program, and proposed a package of consumer measures that would have required a transformation of the solar industry. Woven through this ambitious program was the notion that it was merely a demonstration and not an attempt to accelerate market growth sharply. The result was a wasteful and fundamentally inequitable program design.

Understanding the PUC: An Institutional Analysis

Directly or indirectly most of the problems originated with the commission itself. The PUC is a regulatory agency, accustomed to interacting with business and the public through legal orders and other relatively formal means. The swh program, however, required the arts of persuasion and the skills of entrepreneurship. It was far more like a social-service program run by the Department of Housing and Urban Development than a traditional economic regulatory program. Predictably the ethos of both the PUC and the utilities it regulated was alien to the sort of give-and-take, and informality, that the implementation of social-service programs typically requires.

The contradictory social-service aspect was illustrated by the commission's dealings with the solar industry. In the PUC hearings leading up to the January 1980 decision, participation by the industry was "notably lacking," to use the commission's own words.[15] The absence of the California Solar Energy Industries Association (CalSEIA) and other possible parties to the proceedings was not surprising given that the commission was simply "investigating the feasibility" of solar programs. Nevertheless, the PUC felt confident it could launch a program without advice from a key participant: the industry that could make or break the program. CalSEIA successfully petitioned for a rehearing and argued that credits were a far better marketing tool than utility loans. The commission found this convincing and changed the emphasis of the program from loans to credits.

The damage in this case was not great—a few months delay. The commission's regulatory attitude toward industry performance was more damaging, however. As exemplified by its consumer-protection package, the PUC viewed the swh industry as a new utility that needed tight regulation. Although the industry is young and still has some shady operators, most of the firms are honest. Yet the commission's legitimate distrust of the minority spilled over into adversary relations with the industry as a whole. Furthermore, some PUC staff members displayed a distrust of private business in general. Others have referred to the solar industry as "crooks" and "crybabies" on various occasions.

Commission interaction with utilities also reflected the problems it faced in stepping outside its traditional role. Leonard Grimes, Jr., the commissioner most responsible for the swh program, saw the need for informal meetings with utility executives to get the program moving. Grimes found, however, that the utilities were very concerned over the possible legal ramifications of such ex parte meetings. Perhaps, too, it was due to the regulatory ethos of the PUC that the swh program created a large backlog of inspections to be completed by the utilities. The PUC mandated inspections but did not calculate what would be involved in carrying them out—that was for the regulated utility to worry about. The potential for alienating the utilities, the solar industry, and the public by creating this backlog apparently did not worry the PUC staff and commissioners, who were used to telling people what they must do and then punishing them with criticism or with more-direct financial penalties if they failed.

To the structural incapacities of a regulatory body to execute a service-delivery program there was added a professional indisposition. The PUC staff is dominated by lawyers, accountants, engineers, and managers who have learned the attitudes and jargon of these professions. For all three dominant professions there is a concern for order and precision, a belief in the application of general principles based on reason and experience to particular circumstances or problems, and an inclination to try to use these methods to control and order the world of affairs. Thus, the head of the PUC Energy Conservation Branch, an engineer, almost single-handedly formulated warranty requirements and sought to impose them on the solar industry.

The PUC staff had the power to work its will. The PUC staff, numbering about 900, has generally concerted professionalism and longevity into institutional power autonomous from that of the commissioners. For many years the professional staff decided all utility-rate issues, leaving only the formal ratification to the commissioners themselves.[16] Thus, when the PUC commissioners ordered the staff in December 1980 to use a certain methodology in calculating sizing requirements, the staff ignored the order. In reply to questions by angry members of an advisory committee to the PUC, one staff member replied, "We stand by our original recommendations [to use the staff-recommended methodology] . . . the [December] decision was wishy-washy."

It was the ideological cast of the PUC's policy thinking, however, that caused the most trouble for the swh program. The five commissioners are appointed by the governor, and confirmed by the state senate, for nonoverlapping terms of six years. In 1979 when the PUC commenced hearings on the swh program, all of the commissioners had been appointed by Governor Jerry Brown, and they reflected rather closely his views on the virtues of energy conservation, renewable-energy sources, environmental protection, and the decentralization of economic and political power. They were committed to tight regulation of utility rates in the interests of ratepayers, especially low-income ratepayers. The Brown appointees changed the PUC from an agency that for many years had rather passively accepted the views of the regulated industries to an agency that made a point of challenging them at every turn. The California PUC became known nationwide as one of the most-progressive, reform-minded state PUCs.

Historically the PUC has been relatively invisible to the general public; however, as utility rates skyrocketed and plant siting became controversial in the past decade, the PUC felt increasing pressure from a variety of interests: utility shareholders, business lobbies, environmental organizations, as well as citizens. The PUC can no longer keep its traditional low profile nor can it do what the commissioners determine is abstractly in the public interest without considering the reactions of these groups.

Despite these pressures, the commissioners are probably freer to make policy according to their own judgments than are most other policymakers. This freedom derives from the fact that they are not elected and that they cannot be removed from office by the official who appointed them except for cause. In addition, because commissioners are appointed for six-year terms, which are likely in many case to run beyond those of the governor, most cannot expect to be reappointed. To put it differently, since most PUC commissioners are not expecting reappointment, they do not need to be overly concerned about whom they offend with their policies. As long as the courts can be held at bay, the legislature pacified to the extent that it keeps their budget at a reasonable level, and the interest groups dealt with by a combination of genuine responsiveness and rhetoric, the commission can act as the majority sees fit.

By 1979 this majority was ideologically progressive, favoring ratepayers at the expense of utility investors, being sensitive to environmental concerns, wishing to promote the interests of lower-income groups, and promoting renewable-energy resources and conservation as substitutes for fossil fuels and nuclear power. The shift in commission ideology was apparent as early as 1975 when it adopted lifeline rates. The Brown appointees hit full stride in carrying out their progressive agenda in 1980–1981 as they implemented a rate-of-return incentive linked to a utility's conservation efforts, a zero-interest loan program aimed at conservation, and the DSFP.

The PUC also desired to infuse society with greater rationality, order, and moral consciousness. While the political movement on behalf of state public-utilities regulation originated early in this century primarily in the utilities' desire to forestall local regulation and ownership, there was also an element of scientific management in the movement. This element was distinctly a product of the progressive ideology of men like Woodrow Wilson, Theodore Roosevelt, Robert Moses, Robert LaFollette, and a host of good-government municipal reformers of the period. LaFollette, for instance, wrote in 1913 of the Wisconsin Railroad Commission that it had succeeded in benefiting both "the people" and "the investors" because its regulation was "scientific." The Railroad Commission, he wrote,

> has found out through its engineers, accountants, and statisticians what it actually costs to build and operate the road and utilities. On the other hand, since the Commission knows what it costs, it knows exactly the point below which rates cannot be reduced. It even raises rates when they are below cost, including reasonable profit.[17]

Clearly the California PUC of the late 1970s acted very much in the LaFollette tradition.

One of the great drawbacks of the progressive world view is its reliance on a top-down model of social reform: the experts on top draw up a rational and moral plan to be implemented by the groups, organizations, and individuals in the middle or at the bottom. Give-and-take, partnership, interaction, evolution—these are not themes found much in favor by the progressive world view. Thus, in the case of the DSFP, the utilities were thought to be hidebound, too dependent on conventional energy sources, and still predisposed to the large capital plant expenditures that would expand their rate bases. The solar industry was thought to include a very sizable portion of fly-by-night or incompetent firms. The California Energy Commission, which had devised technical standards and warranty requirements for the solar tax credit, was viewed as undemanding and slow-moving. Consumers could not be trusted to make wise decisions because they were technically unsophisticated and susceptible to the allure of solar technology for its own sake. There is and was at least some truth in all of these beliefs, but the commission concluded that only it could make up for the perceived deficiencies of these other parties, especially in the area of consumer protection. This was surely a dubious inference.

In our view the PUC in effect acted against consumer interests by inducing uneconomic swh installations in homes with gas water heaters. Without a very dramatic, and perhaps far-off, increase in gas prices, these investments will not be beneficial to these consumers. Furthermore, the design standards initially proposed by the PUC were incompatible with swh state of the art as well as with the condition of the solar industry.

Evaluation and Adaptation at the PUC

It is in the nature of things that new ventures will be subject to more mistakes than already institutionalized programs and that new ventures launched by enthusiasts are bound to be more laced with implementation problems. This is not to say that enthusiastic new ventures should not be undertaken. Often the problems are simply the price we must pay to overcome an equally frustrating level of social inertia. But it is surely desirable that new and problematic ventures, especially in public policy, be accompanied by systematic efforts to detect problems as they emerge and devise remedies for them. With regard to the DSFP, this process did occur in a limited way.

Most importantly, whatever resources were unwisely committed to gas retrofits in the end, they were far less than those contemplated by the original planners. The PUC report "Financing the Solar Transition," prepared in 1979, studied the possibility of reaching 20 or 80 percent of the residential water-heating market over a ten-year period, though it added that the PUC was not recommending these targets at the time.[18] By the time the PUC issued its DSFP order in January 1980, however, the targets were reduced to 2 percent of the gas water-heating market and 10 percent of the electric water-heating market. In September 1981, the PUC set a target of 1 percent specifically for the single-family gas market, increased the electric water-heating target to 15 percent, and added a separate target for the multifamily market.

In a series of decisions beginning in December 1980, the commission modified a number of consumer-protection measures that otherwise might have crippled the entire program. The PUC backed down on its proposal to require warranty responsibilities up to ten years. It eventually adopted warranty requirements identical to those required to claim the solar tax credit (a three-year warranty on collectors and tanks and a one-year warranty on other parts). Prescriptive standards went through a number of changes between December 1980 and June 1981.[19] The final set of standards is widely acknowledged as offering substantial protection to the consumer yet retaining reasonable flexibility from the industry's point of view.

The commission also had second thoughts about the low-interest loan program. In July 1981, it cancelled the loan program for one utility, in part because of its high cost. In the remaining utility's loan program, strong evidence of wildly inflated swh prices appeared within weeks after the utility began lending. The PUC responded quickly with an emergency order placing a ceiling of $4,000 on the amount of money an swh buyer could borrow from the utility. This order was later modified to allow borrowers to obtain additional amounts at roughly the market interest rate.

A final example of adaptation concerns the growth of the PUC's concern for solid technical and economic analysis. In September 1981, such analysis was just one of over a dozen tasks the commission instructed its staff to undertake as part of the evaluation of the DSFP. By late 1981, several key staff had convinced the PUC to make solar-system monitoring the centerpiece of the evaluation effort. The result will be the most-extensive testing of swh performance ever undertaken; the data collected will be invaluable in accurately evaluating the economics of swh.

There are two quite different ways to look at the PUC's efforts in diagnosing and correcting its errors. One is to concentrate on how long it took and how modest the improvements have been and promise to be; the other is to marvel that a large and complex organization such as the PUC has been able to start the learning process at all. Our preference is to take the second viewpoint. The implementation literature is full of gloomy accounts of how things go wrong, and indeed our own account to this point has accentuated the negative. In this context, the few bright spots in the history of policy implementation experiences ought to be magnified and brought into sharper focus.

Why was the PUC able to correct some of its mistakes? Part of the explanation lies in conflicts between staff members. Newer staff were more inclined to launch a large and costly solar program for ideological reasons. Traditional old-line staff were concerned primarily with preventing the waste of ratepayer funds. Old-line staff tried to modify the program in light of their primary goal of protecting ratepayer interests. This is illustrated by the progressive scaling back of the program, the heightened priority of solar-system testing, and the modifications of the utility loan program. As one of the older staff told us, it was "quite an accomplishment" to have scaled the gas program down to the point where the overall program at least appeared to break even.

The impetus to modify PUC consumer-protection measures had to come from outside the commission. At the same time that it launched the program, the PUC also established the DSFP Advisory Committee, made up of representatives from the utilities, the solar industry, environmental groups, and government bodies outside the PUC. Each organization had some stake in the outcome of the program. The PUC directed the commission to make recommendations on solving implementation problems that would arise in the DSFP.

The advisory committee was primarily responsible for transforming the PUC's original, overly ambitious consumer-protection proposals into a workable program. From January 1980 through the fall, the commission relied mainly on the staff from its Energy Conservation Branch to develop swh design, performance, and warranty standards. During this period the advisory committee was disorganized and largely disconnected from the standards-development process. The committee submitted letters to the commission concerning the standards and warranties proposed by the Energy Conservation Branch staff; however, the committee developed no clear communications channel with the commission or the staff, and the letters had no apparent effect. Once the program was launched, the PUC felt pressure from the industry and utilities, as well as from the advisory committee, and slowly realized that the committee was a greater source of expertise on solar energy than its own staff. The advisory committee also became more aggressive, successfully demanding that the commission designate a member of Commissioner Leonard Grimes's personal staff to be the continuing liaison between the commission and the committee.

In December 1980, the PUC eliminated the warranty provisions extending beyond the fifth year, and in March 1981 it scrapped its warranty proposals altogether, adopting warranties identical to those required by the Energy Commission in conjunction with the solar tax credit. The advisory committee can take a good deal of credit for the evolution of the warranty requirements and the prescriptive standards as well. By June 1981, the rigidity of the 1980 proposed standards had been largely eliminated; the standards emerging at that time (and still emerging) represented a consensus among technically knowledgeable people from the various affected sectors. The consensus was not perfect, but it was not more strained than is common in the general practice of writing industry-wide consensus standards.

A second outside impetus for change in the consumer-protection package came from CalSEIA, the solar trade association. CalSEIA strongly opposed the commission's initial warranty proposal and initiated a lawsuit challenging the commission's authority over such matters. As the suit was slowly moving toward the state supreme court, the commission backed down and CalSEIA dropped the case.

As finally implemented, the DSFP was a far better program, yet it still suffered fundamental flaws. We turn now to an examination of implementation problems in two other policies that encourage solar water heating: the solar income-tax credit and local solar mandates. Together with the PUC program, they form the three most-prominent government programs in California aimed at encouraging swh. A comparison of the three will lead us to conclusions regarding which is the superior policy design.

The Solar Income-Tax Credit

The PUC program is a subsidy financed by utility ratepayers and administered by the utilities in conjunction with the PUC. A comparable program of subsidies financed by state taxpayers and administered by the Franchise Tax Board in conjunction with the California Energy Commission (CEC) was inaugurated four years earlier. It too has had implementation problems, some of them similar to those that beset the PUC program.

Current provisions of the California solar income-tax credit allow home owners to claim a 55 percent credit if they purchase any of several renewable-energy systems. (Here we focus only on swh systems.) A claimant must deduct the 40 percent federal solar-tax credit from the state credit and then can deduct the remaining state credit directly from the tax bill. In 1979, the most-recent year for which data are available, California home owners installing swh claimed about $5.6 million in state tax credits, with about half this amount being carried forward to future tax years.[20]

When first considering creation of a tax credit, the legislature paid scant attention to swh cost-effectiveness and its relation to the type of backup energy source. Consequently the tax credit did not distinguish between systems that have electric rather than gas backup. Subsequently the California Energy Commission (CEC) produced various studies of swh economics. Prior to the departure in 1980 of CEC Commissioner Ronald Doctor, these studies showed swh to be economical even when supplemented by gas. There were a number of serious analytic flaws in these studies, however.[21] There is now a fair amount of dissension within the CEC over the cost-effectiveness of swh with gas backup. We agree with the skeptics.

The CEC did not try as hard as the PUC to develop a consumer-protection component for the tax-credit policy. Over the years, the CEC has created various programs and written regulations designed to ensure the quality of swh installations. The package has never quite worked. Since 1977, the CEC has conducted a program that tests both the durability and thermal efficiency of solar collectors. The main defect in the program is that collector efficiency is an incomplete measure of system performance. It is a difficult though necessary task to test an entire system rather than merely one component. In contrast to the PUC, the CEC has also neglected to perform any extensive monitoring of installed systems.

The CEC requires that tax-credit claimants employ a licensed solar contractor. The licensing requirements are so lax, however, that consumer protection may actually be harmed by this provision. Currently a person holding a contractor's license in a related field (for example, plumbing, electrical work, or general building) can obtain a solar license by paying a fee. To the potential swh buyer, the solar license may create the illusion that the contractor is competent to install solar equipment. A meaningless solar license may make a buyer less inclined to investigate a contractor's true credentials and past work record.[22]

The Energy Commission is also empowered to issue prescriptive standards that determine which systems qualify for the tax credit. The CEC has performed poorly in this role. Current prescriptive standards require little more than that the collectors face south and that certain energy-conservation measures be undertaken.[23]

Even if the CEC had issued stronger standards, their effectiveness might have been limited. Neither the Energy Commission nor the State Franchise Tax Board has the resources to verify the accuracy of information submitted by taxpayers, much less conduct inspections of the actual equipment. The embryonic CalSEAL program (a joint CEC and CalSEIA effort) is the only significant step toward meaningful enforcement of consumer-protection measures.[24]

The state did not devote more funds to enforcement for several reasons. First, promotion is cheap but enforcement is expensive. Second, promotion, especially subsidies, wins friends while enforcement can create enemies. The PUC's ability to dispatch armies of utility inspectors to enforce its consumer-protection measures reflects the commission's relative insulation from the political forces that dominate solar policy in Sacramento. By contrast, neither the CEC nor the legislature was willing to spend several million taxpayer dollars on enforcement.

Equity is another problem that has plagued the implementation of the tax credit. Persons in relatively high income brackets have consistently claimed most of the tax credits. In 1980, the legislature amended the law to allow lower-income persons to participate. The amendments provided that persons with adjusted gross incomes of less than $15,000 could receive a tax credit in excess of their tax liability. In effect, the state would write a check for the difference. A year afterward, however, the unexpectedly high cost of this provision led the legislature to delete the provision.

Solar Mandates at the Local Level

More than a dozen California cities and counties have passed ordinances requiring swh in new residential construction, among them, the counties of San Diego, Santa Barbara, and Santa Clara, and the city of Sunnyvale. In one city, Davis, the town planning commission requires swh without reference to an ordinance. In one county, Santa Clara, the local ordinance also applies (after 1983) to existing housing: swh must be retrofitted at the time the dwelling is sold. The main promoters of the idea have been the CEC, Governor Brown's SolarCal Council, and the Campaign for Economic Democracy led by Tom Hayden. Part of the motivation behind the movement on behalf of solar mandates was the belief that the tax credit was not working fast enough. Some political observers believe that part of the motivation sprang from the belief that solar mandates could be a catalyst for the creation of a grass-roots political organization.

Generally the local ordinances fail to distinguish between homes with gas and electric water heating. Santa Barbara County is the only exception in that it requires swh only for new homes with electric backup. Either the designers of the local ordinances believed the CEC and PUC contentions that swh was—or very soon would be—cost-effective or they did not care. As one member of the Davis Planning Commission replied when asked whether the commission had examined the economics of solar water heating: "No. We know solar makes sense. You can do a cost-effective study to show anything. The public doesn't care about cost-effectiveness; the industry doesn't care about cost-effectiveness; only government cares about cost-effectiveness."[25]

The designers of local solar ordinances have generally slighted consumer protection. San Diego County is the only local jurisdiction that has developed sizing and design standards. Santa Clara County enforces a 50 percent minimum sizing guideline. The Davis Planning Commission exercises some quality control over systems installed on a case-by-case basis, as does Santa Barbara's senior energy specialist. Overall, however, localities have added few consumer-protection measures to those existing under the tax credit.

Building inspectors are responsible for enforcing any regulations specific to the solar mandate. They also enforce existing building, plumbing, and electrical codes that can have some impact on the quality of a solar system. These codes can help ensure good workmanship in the installation, such as a properly sealed roof and safe electrical wiring, but the codes do not currently address performance. As one Davis building inspector said, "We can make sure that it's put in right, but we can't tell if it works." In any case, not all building inspectors have the expertise to tell whether "it's put in right," and local building departments do not ordinarily have the resources, given the revenue-limiting effects of Proposition 13, to hire solar specialists or to train their existing inspection force.

Comparing Programs

Government has had difficulty encouraging solar water heating in an effective and efficient manner. We trace this difficulty to three characteristics of swh. First, solar advocates have oversold the technology. The promised benefits include cheaper energy, more jobs, a cleaner environment, financially more healthy utilities—the list goes on. The fact remains that the sun is a fairly expensive means of heating water. Although electricity is even more expensive, natural gas is still much cheaper. The ideological aura surrounding swh can cloud the judgment of public officials.

Second, even when officials are able and willing to take a critical look at costs, they may find it hard to do so. The technical issues are complex, and the pitfalls in the way of competent cost-effectiveness analysis are numerous. Such analyses are particularly sensitive to assumptions about future gas prices, swh system reliability and performance, and consumers' hot-water consumption.

Finally, implementation problems are found to arise in any solar program because of the desirability of including a consumer-protection feature. Government cannot merely stimulate demand if it is to achieve the societal benefits contemplated. The swh industry is young, and consumers are not well informed. Without effective monitoring programs, the quality of the systems installed will not be adequate to deliver the energy savings projected.

The alternatives to swh appear even more attractive in light of these characteristics. Electric water heaters can be converted to gas, and heat-pump boosters added to electric water heaters are also promising. If government still insists on launching a program to encourage swh, however, we believe that a PUC-administered and ratepayer-financed program represents the best approach of the three we have examined.

Any one of these policies alone could conceivably encourage swh with electric backup in both new and existing buildings. Combinations of policies are also possible.

In the following discussion, we list a number of characteristics that a solar program should display if it is to avoid implementation problems. The PUC program has more of these characteristics than either the tax credit or local mandates and hence is a preferred approach to encouraging solar water heating.

Technocratic Decision Making

A solar program should be technocratic in nature. The government body administering the program should be capable of dealing with complex technical and economic questions. It should be willing to limit program applicability to electric water heaters and institute a meaningful consumer-protection program. Both of these measures will likely antagonize certain groups in the implementation process; hence the agency should enjoy a degree of political insulation that allows it to carry out these measures.

A state PUC is probably in the best position to make the technocratic decisions necessary for a well-designed solar program. The key characteristics of a PUC in this context are its relative insulation from the political forces that buffet a state legislature, a state energy, or resources agency, and local government bodies and its broad powers to regulate utilities without detailed oversight or obstruction by other branches of government.

In California the PUC was powerful enough and politically insulated enough to institute demanding consumer-protection measures. Although the commission was not so technocratic as to limit the DSFP to electric retrofits, it did distinguish the relative cost-effectiveness of gas versus electric retrofits, and it did reflect this distinction in its policy design. Such a distinction was lacking in the actions of the legislature, the Energy Commission, and all localities except Santa Barbara County. The legislature and local governments tended to view solar energy in more-ideological terms. Many CEC staff shared this tendency, and in addition, the Energy Commission was much more accountable to the legislature than was the PUC.

Adaptive Capability

Another positive trait of an organization promoting swh is the ability to respond relatively quickly to unforeseen implementation problems. One would expect administrative agencies to be able to respond to implementation problems with greater speed than more-deliberative bodies such as the legislatures, city councils, or county boards.

This notion held true in the California experience. At times the PUC behaves like a plodding, bureaucratic organization, but it also has the power and capability to act quickly. The commission's emergency order placing a ceiling on utility loans is a good example of this. The legislature did not grant the CEC the broad powers to administer the tax credit that the PUC had over its program (as derived from the commission's general powers). Had it been given more discretionary power, however, the CEC probably could have responded as quickly as the PUC. It also easily matches the PUC in technical and economic expertise, an area where localities are clearly disadvantaged.

Resources

The capacity of an agency to devote resources to consumer protection is also of interest. Consumer protection is costly when inspectors are needed to ensure proper installation. An agency should be able to draw on ample funds for this purpose. Ideally, it would also have a cost advantage in providing consumer protection by controlling a source of potential inspectors (for example, existing staff performing similar tasks).

The PUC clearly had the needed capabilities. It could easily direct the spending of millions of dollars in ratepayer's funds, using utilities as intermediaries. Further, the utilities already had staff in the field to serve as inspectors. At a time when budgets are extremely tight for most state and local agencies, a public utility commission's subtle taxing power (through rate making) is an enormous advantage. Other administrative agencies may have neither the funds nor the inspectors to implement a consumer-protection program on the same scale as a PUC. Localities have building inspectors but generally lack the funds to train their existing inspectors or hire solar specialists. The subtle taxing power of a PUC is an advantage, but it also poses the risk of wasting ratepayers' money.

The California PUC was not thoroughly technocratic in its program design. If program goals are met, utilities will spend $83 million in subsidies to gas retrofits, causing a net loss to ratepayers of $42 million (net present value over the life of the program). In contrast, tax-credit claims for swh systems totaled only $2.8 million in 1979, with an additional $2.8 million carried forward to future tax years.[27] A public utility commission can flex a lot of financial muscle, for better or worse.

Equity and Fairness

A good program should be fair and equitable. The central equity question concerns the treatment of ratepayers who do not receive program subsidies. A genuinely cost-effective subsidy program would leave nonparticipating ratepayers better off or at least unharmed. A second equity concern is with the distribution of subsidies among the pool of potential beneficiaries. If the private benefits of receiving the subsidy are large, this equity issue could loom as quite important.

If we are right in thinking that a state PUC is relatively better able than other governmental agencies to implement a technically and economically sensible program, then a PUC-run solar program is more likely to be equitable than a tax credit or local mandate. Yet in the California case, the DSFP was in fact not designed to be cost-effective. Ironically, this occurred because of the PUC's desire to attend to the second equity concern. That is, the PUC sought to distribute the supposed private benefits of the solar subsidy to all classes of ratepayers, including those with gas water heaters. The irony is compounded because the private benefits of participating in the subsidy program were, for these recipients, actually negative.[28]

A third equity consideration is the choice of a base for financing the solar program. The cost of a local mandate requiring solar water heating in new homes is borne largely by new-home buyers. As a group, these individuals are probably better off than the population at large. In our view, however, it is distinctly unfair to impose the costs of any swh program on a narrow and arbitrarily delimited group of citizens such as buyers of new homes. The benefits of the program presumably accrue to all persons, and the method of financing the program should reflect this. The unfairness of imposing the costs on home buyers is substantially lessened if the buyer stands to benefit financially from the swh investment. Without a subsidy, however, swh is not cost-effective from the private point of view, even for electric water heating.

Regarding the choice between ratepayer and taxpayer financing of the solar subsidies, we once again favor the approach involving a PUC. Ideally, a solar program would be cost-effective and financed with public monies from a progressive funding mechanism (a progressive income tax). We have argued, however, that a taxpayer-funded subsidy has the lesser chance of being cost-effective. Ratepayer financing seems to be the next best choice. Assuming that a PUC program has net benefits to nonparticipants and that the incidence of the benefits is the same as the incidence of costs to ratepayers, then a PUC program should have neutral income-distribution effects. Nevertheless, there is risk in this approach in that a program that wastes ratepayers' money may have regressive distributional effects.

From the standpoint of fairness, the involuntary nature of local solar ordinances should be counted heavily against such an approach. There are significant financial risks associated with the purchase of an swh system. There may be chronic mechanical problems, and energy prices might not rise fast enough to warrant the investment even if the buyer uses a very low discount rate. If there were large and clearly defined social benefits to a compulsory program and if there were no alternative way to spread the financial burden more broadly, perhaps local mandates could be justified. However, we have seen that neither of these conditions actually holds.[29]

The local regulatory approach could be uncoupled from the implicit requirement that the financing be undertaken by home buyers. The building developer or contractor could be responsible for complying with the ordinance, but utility financing or a tax credit could help subsidize compliance. Indeed, this two-pronged approach is precisely what a new home buyer faces in those localities in California that have passed an ordinance.

Summary and Conclusion

Solar water heating is a cost-beneficial substitute for electric water heating from a social point of view; however, for reasons having to do with the way electric power is priced to consumers, the privately borne costs of electricity use are likely to be lower than the social costs. Hence there are inadequate incentives for consumers to substitute swh for electric water heating. A public policy designed to encourage swh may therefore be in order.

Three different methods of creating and administering such a policy have been used in California during the past several years. One method is taxpayer financing through tax credits, with the administration of the program handled jointly by the tax-collecting agency and an energy agency in the executive branch. A second method is financing by utility ratepayers, with program management the responsibility of the state PUC in conjunction with the utilities. The third method is local ordinances mandating swh installations in new residential construction. This method implicitly relies on financing by home buyers and the administration of the standards for swh performance by local building departments.

With respect to criteria of efficiency and equity, local ordinances are likely to be inferior to the other two policy approaches. Between the tax-credit approach and the PUC approach, the latter is probably superior. Our judgment is based on considerations of which administering institutions are more likely to avoid certain predictable pitfalls of the implementation process. The main pitfalls to avoid are extending the program to cover gas, as opposed to electrical retrofits, and failure to create effective and reasonable consumer-protection measures. The PUC approach is superior to the tax credit primarily because a PUC is more likely than the legislature to have access to the technical and the economic expertise to identify these pitfalls and because its relative insulation from the political process permits it the latitude to act in accordance with its expertise.

A PUC is not guaranteed to act wisely in any absolute sense. In the case of California's PUC program, the commission did not perform well at all, though the DSFP was a better program than the solar-tax credit and local ordinances. This fact underlines the sensitivity of policy that is only conceptually desirable to the vagaries of the implementation process it must undergo in the real world. It further underlines the fact that subsidizing the conversion of electric water heaters to gas and installing heat pumps to boost the output of those that cannot be converted are even more cost-effective, and hence superior, policies to retrofitting with solar.

Notes

1. In addition, conventional energy sources often impose environmental costs of various sorts. If these costs appeared in the prices paid by consumers, they would be another reason to invest in swh.

2. Robert Foster, "Solar Energy in the 1980s" (San Francisco: California Foundation on the Environment and the Economy, 1981), p. 16.

3. Ibid. These societal cost estimates are roughly compatible with those made in: Solar Energy Research Institute (SERI), *A New Prosperity: Building a Sustainable Future* (Andover, Mass. Brick House Publishing, 1981). The SERI report offered the following estimates for the current price of hot water by source:

Source	1978 S/kilowatt hours
Electric water heating	0.030–0.070
swh without 40 percent tax credit	0.027–0.094
swh with 40 percent tax credit	0.016–0.059
Heat pump	0.034–0.045

	1978 S/million Btu
Gas water heating	3.50–10.00
swh without 40 percent tax credit	8.00–27.50
swh with 40 percent tax credit	4.75–17.30

4. Foster, "Solar Energy in the 1980s," p. 15. Recent staff studies performed at the California Energy Commission estimate life-cycle swh costs at about $5,600. CEC, Hearings on Residential Building Standards, Sacramento, California, March 18, 1981. "Solar Energy in the 1980s" put the cost at $6,300 to $7,300. One could also argue that swh should be subsidized when displacing gas because solar energy is environmentally benign or because gas production

is subsidized; however, gas is a relatively benign fuel from an environmental viewpoint, and subsidies for gas production are relatively small compared to those enjoyed by nuclear power. We seriously doubt that these considerations outweigh $2,000 or more in added costs resulting from each swh installation where gas is the backup fuel.

5. Ibid., pp. 15–16; SERI, *A New Prosperity*, pp. 72–73; John Rothchild, *Stop Burning Your Money: The Intelligent Homeowner's Guide to Household Energy Savings* (New York: Random House, 1981), pp. 150–163.

6. Eugene Bardach, *The Implementation Game* (Cambridge: MIT Press, 1977), pp. 85–90.

7. PUC Decision 92251, September 16, 1980.

8. CEC, *Analysis of the California Solar Tax Credit 1978 Returns* (August 1980), p. 46.

9. The court held that the commission could not order such a program without specific legislative authorization. The general-powers clause in the PUC code was insufficient grounds. Two years later, however, the court refused to review the PUC's decision establishing the solar program, though the commission lacked explicit authorization.

10. PUC Decision 92251, p. 5.

11. This conclusion is based on a comparison of the private water-heating cost estimates in Foster, "Solar Energy in the 1980s," p. 16, after deducting a $920 PUC subsidy from private swh costs (electric backup).

12. This conclusion is based on a comparison of the private water-heating cost estimates of ibid. after deducting a $760 PUC subsidy from the private swh costs (gas backup).

13. PUC, *Financing the Solar Transition*, January 2, 1980, p. 16.

14. The most-ambitious such program was the California Solar Energy Approval Label (CalSEAL). This program was a joint effort by the CEC and the solar industry to give consumers some assurance that a swh system met the requirements of the solar income-tax credit. CalSEAL had limited funding and scope when the PUC was contemplating its solar program. CalSEAL principally aimed to certify that participating businesses met tax-credit warranty requirements. Despite the certain doubts about the effectiveness of CalSEAL, the Energy Commission saw it as a valuable, albeit embryonic, program. The CEC believed that CalSEAL could be expanded to become an independent, nonprofit corporation with duties such as standard setting, system testing, and spot inspections. Such a program would be a form of industry self-policing that is common in many mature industries.

A program such as CalSEAL is a long-term solution to the consumer-protection quandary. CalSEAL may have to start from scratch after the PUC program ends. Consumers would likely have been better off if the PUC had tried to integrate its consumer-protection plan with CalSEAL.

15. PUC Decision 92251, p. 17.

16. A. Janell Anderson, "Decisionmaking in the California PUC: Cal-

culating Rates of Return," California Government Series II, No. 10 (Davis: Institute of Governmental Affairs, University of California at Davis, February 1979), p. 16.

17. Quoted in Douglas D. Anderson, *Regulatory Politics and Electric Utilities: A Case Study in Political Economy* (Boston: Auburn House, 1981), p. 55.

18. PUC, *Financing,* p. 67.

19. See PUC Decisions 92501 (Dec. 5, 1980) and 92769 (March 3, 1981), and Commission Resolution No. EC-11 (June 2, 1981).

20. CEC, "Analysis of the California Solar Tax Credit 1979 Returns," draft, table 4.

21. See, for example, the CEC Staff Report, "Solar vs. Conventional Residential Water Heating" (August 1980). The report made two key assumptions that tend to inflate the total amount of gas consumed, hence the energy savings of the swh systems. The report assumed a four-person household (three is the modal family size in California), and it assumed that an old, relatively inefficient water heater was being installed in a new home. In addition, the report did not account for the electricity use of the pump in the solar system. Furthermore, the report used a rather low estimate for the cost of the solar system and then added the cost of the conventional system before calculating the tax credit.

22. The consensus among several staff members at the CEC, the Contractors State Licensing Board (CSLB), and CalSEIA is that the consumer would be better off without the current licensing procedure. The CSLB is currently developing tougher licensing requirements.

23. The CEC is also in the process of upgrading its prescriptive standards. Interestingly, the PUC's standards served as a starting point for the Energy Commission.

24. See Footnote 14.

25. Interview with Davis Planning Commission member, Davis, California, February 4, 1982.

26. Authors' survey of building officials and energy staff in localities where a solar ordinance has been passed.

27. CEC, "Analysis of the California Solar Tax Credit 1979 Returns," table 4.

28. Assuming that the private costs of swh minus the subsidy were still greater than the private costs of gas water heating.

29. We are not necessarily opposed to all mandates regarding energy use. We distinguish between risky and virtually risk-free investments forced on the consumer. A properly designed appliance-efficiency standard may impose an incremental cost with a nearly certain payback period of several years. This is far less risky to the consumer than forcing a swh purchase. Solar systems gen-

erally have longer payback periods, and their attractiveness as an investment is more dependent on forecasted increases in energy prices. The systems also carry a greater risk of mechanical failure than other energy-saving devices. By this reasoning, we favor the two voluntary programs (DSFP and tax credit) over local solar mandates.

5

Suburban Resistance to Density Increases near Transit Stations in the San Francisco Bay Area

Robert A. Johnston and
Steve Tracy

Land-use intensification yields significant savings in energy conservation, both within structures and in transportation. Medium- and high-density residential developments can take advantage of solar heating and wind cooling and can conserve energy as effectively as can the much more widely accepted low-density designs (Lovins and Lovins 1980, pp. 96–98).[1] High densities also permit the use of district heating and cogeneration. Combined with site design and community layout to facilitate pedestrian, bicycle, and bus access modes, medium-density developments can result in substantial energy savings in transportation when compared to typical single-family subdivisions (U.S. GAO 1981, pp. 16–17).

Keyes and Peterson (1980) claim that a strong national shift toward multifamily residential construction could save 0.5 quadrillion Btu's for heating and cooling and about 1 quadrillion Btu's in transportation annually by the year 2000. They point out that urban per-capita gasoline consumption increases with total population, the proportion not living in high-density areas (over 10,000 people per square mile), and the proportion of jobs located in the central business district (U.S. GAO 1981, pp. 43–44).

Keyes (1976) points out that national mandatory auto fuel-efficiency standards, gasoline tax increases, and thermal-efficiency requirements for new buildings would have larger effects (over 4 quadrillion Btu's saved). The Solar Energy Research Institute (1981) claims much higher savings from these three measures. Keyes states that the building standards would decrease the differential for heating and cooling between low-density and high-density buildings. Efficient autos would also reduce the relative benefit of compact development. However, he notes that gasoline real-price increases will tend to concentrate growth. We maintain that the phasing out of federal support for freeway and sewer building, rising real land costs on the urban fringe, and the decreasing real, after-tax incomes of American households will decrease the percentage of low-density suburban development in the next twenty years.

Because thermal building efficiency has been well recorded (Solar Energy Research Institute 1981), we chose to investigate the relationships between land use and transportation. In order to perform empirical research, we examined land uses affecting an existing heavy-rail system, San Francisco Bay Area Rapid Transit (BART). BART suffers from a lack of countercommuting because it primarily serves one central business district, San Francisco, and serves low-density suburbs with poor ridership.

Background

People go where transportation is available. Historically the size of our cities has increased as our ability to move about has grown with the shift from walking to horse and cart to mechanized travel. Faster transportation has been a cause not only of size increases for cities themselves but also for supplying ever-larger populations within the cities from larger surrounding areas. The size and density of a city is chiefly determined by the modes used by its residents to move around as they go about their lives.

In this century in America, particularly after World War II, the widespread use of the automobile allowed settlement to occur over wide areas of land. This transportation method was encouraged by devices like tax write-offs for auto loan interest, low gasoline taxes (less than 10 percent of what is common in Western Europe), controlled crude-oil prices, and a heavy government commitment to highway construction (Stobaugh and Yergin 1979).

This increase in mobility allowed expansion out into the suburbs. Various government policies subsidized low-density, single-family home ownership. The chief incentives were the federal tax deduction for home-mortgage interest and FHA mortgage-loan guarantees. Federal financing of municipal waste-treatment plants also encouraged low-density development. Although this suburbanization is certainly beneficial to most of the individuals involved, there are many public and private costs caused by this low-density pattern: the premature urbanization of prime agricultural land, high costs for public services, higher regional air pollution and energy conversion, and poor levels of service for the transportation disadvantaged (Council on Environmental Quality 1974).

As we experience rising real liquid-fuel and highway costs, the burden on the individual and our society of maintaining America's existing automobile-dominated transportation network is increasing. Although transit ridership has been declining for many years as a proportion of total and of urban trips, local and national concern over transit development is increasing.

Literature Review

Energy analysts disagree as to whether transit systems save energy, compared to auto transportation. Lave (1977) states that BART saves energy per passenger mile (versus autos) in operation but that due to the much greater energy expenditure in system construction (versus freeways), the breakeven point for overall energy use is around 500 years after operations began. Lave maintains that buses use less energy per passenger mile than other modes and advocates their increased use, as well as improvements in auto efficiency. Hannon (1977) argues from national data that rail systems are more energy efficient than buses and cars.

Hirsch (1975) claims that urban buses and trains are more energy efficient than autos but states that increased use of mass transit cannot decrease urban energy conversion much because mass transit accounts for such a small share of urban trips (2 to 3 percent); only about half of the new transit riders would have driven autos because the others were auto passengers, pedestrians, and bus riders; expanded route coverage and more-frequent service lowers train loading factors; and autos usually are used to get to the train system. For immediate energy-conservation results, he recommends gasoline taxes, and for medium-term energy savings, new auto-efficiency standards. He also advocates urban transit improvements, for longer-term energy conservation.

One should note, though, that rail transit uses electricity, not petroleum (Craig 1976), and dependence on foreign petroleum is the major energy problem for the United States. Brown, Flavin, and Norman (1979) emphasize that as petroleum becomes more expensive, auto use increasingly will be seen as profligate, compared to food production and home heating.

Peskin and Schofer (1977) have constructed an urban simulation model that shows that a multicentered urban pattern, one with several dense nodes served by transit, is the most energy efficient. The efficiency results from the cross-commuting pattern among centers. Commuting to a single central business district (CBD) from low-density suburbs results in nearly empty cars and buses in the countercommute direction. Since the cars start out empty in the morning and fill up on the way into the CBD, and then return nearly empty, the rush-hour capacity factor is around 25 percent.

Pushkarev and Zupan (1977) identify the levels of residential density necessary to support different types of transit from empirical research in the United States. For example, an overall urban density of seven units per acre may work effectively with a lighter type of transit such as dial-a-ride or small buses but will not provide the number of riders needed for a more-costly fixed-rail system like BART. This is true even if these densities are in corridors, and there is a large, dense downtown to serve as a magnet for riders.

Another point made by Pushkarev and Zupan is the strong attraction an established downtown has to transit riders. A given level of commercial development at an outlying node will not bring as many riders as will an equal amount of new development in an existing large CBD. So the transit system must add costly extra capacity (which will have a peak-use load factor of only 25 percent) to accommodate new passengers going downtown when it could carry those same passengers at little additional cost to work near an outlying station, since they would be riding nearly-empty cars counter to the main flow.

Furthermore, according to Pushkarev and Zupan, a given percentage increase in residential density within one-half mile of a transit stop or station brings about a greater percentage increase in transit ridership than if the new development were farther away, holding constant all other factors such as income, size of family, and automobile ownership. From their tables, we can estimate that in the Bay Area suburbs served by BART, this density-to-ridership increase ratio for growth within one-half mile of stations would be about 1 to 1.2.

Dunn (1981) performed a comparative analysis of European and American air, rail, auto, and public transportation. The factors that have contributed to transit's success in West Germany are higher-density development, less suburbanization, fewer cars (290 versus 500 per 1,000 population, or about the level the United States had in 1950), more government coordination and flexibility, a longer history of public ownership of transit systems, higher past use rates and political support, and higher gasoline taxes (more tax money per gallon is earmarked for transit in Germany than the total 4 cent per gallon federal gasoline tax). European countries have high auto sales taxes, by value and engine displacement, which reduce auto use. Dunn calls for increased coordination of transit and land-use planning in the United States. He cites an Office of Technology Assessment report suggesting tying sewer and water grants to transit and zoning.

Several studies examine the effects of transit-system development on land uses. Knight and Trygg (1975), in a study of American and Canadian transit systems, concluded that transit development can result in land-use intensification but only when demand, economic strength, zoning, community acceptance, and other factors are all favorable, and even then the changes will take place slowly.

The Urban Land Institute (1979) examines several successful efforts at shared development, use, and ownership of buildings near train stations in North America. The study demonstrates how transit agencies can help to bring about more-intensive land use, commercial or residential, that will in turn benefit transit by attracting riders. This can be done by providing land for development over underground transit facilities or by using excess land acquired during construction. The transit agency may share, lease, or sell the land to be developed by private firms.

The discussion of Park Place, a development on one of Toronto's new subway lines, is of particular interest. Unlike the Bay Area, Toronto employs a high degree of coordination between land use and transit planning. Yet nearby home owners impeded land assembly for this development so long that all of the intended development was not completed.

In the BART Impact Program Final Report prepared by the Metropolitan Transportation Commission (1978), the authors perform an analysis of why

BART failed to bring about land-use intensification. The reasons they state are that transit availability is only one of the factors the individual or firm considers when making a location decision, urban blight impedes development around some of the stations, and BART's lack of redevelopment and condemnation authority prevents system officials from taking a direct role in bringing about higher-density development. This report states that BART's existence is evidence of public support for a strong downtown San Francisco, which will tend to reduce sprawl. Another Metropolitan Transportation Commission Study (1979) looks at the BART-caused impacts in the local communities. Each section is begun by showing how pre-BART expectations differed from post-BART experience. The section on local land-use policy points out the desirability of not only allowing intensive developments near stations but also of preventing similar developments elsewhere. This report points out that local transit-reinforcing land-use policies and implementation strategies must be worked out before a transit system is built in order to increase system efficiency.

Webber (1976) demonstrates how the original BART planners' misconceptions about mass transit led to an expensive system that does not meet the needs of many commuters. He points out that the same number of passengers could be carried on a fleet of buses, the cost of which would be less than one-half of one year's payment on the thirty-year construction bonds for BART ($40 million versus $1.6 billion). He asserts that BART actually contributes to sprawl by reducing commuting times from the outermost low-density suburbs to San Francisco.

BART History

Over thirty years ago Bay Area leaders recognized that the carrying capacity of the bridge and highway system bringing automobile-driving commuters into San Francisco was finite. The expense and difficulty of providing daytime parking for all of the automobiles was also seen as placing a limit on the number of suburban residents who could be employed in the city. To Stephen Bechtel and other members of the Bay Area Council, a kind of high-level chamber of commerce and political-support group, these limits threatened a shift in growth and power away from San Francisco to other areas in the region with better access. Being growth-oriented entrepreneurs, these members of the San Francisco business community sought an engineering solution to bring more workers into the city, leaving their cars and attendant pollution and congestion behind.

By connecting the central districts of the Bay Area's two largest cities (San Francisco and Oakland), several smaller cities (Richmond, Berkeley, Hayward, and Fremont), and several growing suburbs, BART was to make an enormous area accessible for a small fee. Voters were told it would be possible to live, work, shop, and play in a variety of settings within brief walking distance of the stations. The advantages of having residential, office, and commercial space available near transit were expected to spur a boom in land sales and construction near the stations. Land use was expected to intensify as demand for well-located parcels rose. The islands of high-rise commercial and residential development, which had sprung up around stations in Toronto's newly constructed subway lines, were cited as examples of what would happen in the Bay Area (*Oakland Tribune,* August 6, 1967).

After fifteen years of planning and politicking, which saw Marin, San Mateo, and Santa Clara counties drop out of the transit district, voters in San Francisco, Alameda, and Contra Costa counties in 1962 approved a bond issue to finance construction of a seventy-two-mile fixed-rail transit system with thirty-four stations serving fifteen communities in the three counties. Before construction was completed, the initial allotment of funds had been exhausted, and additional sources were sought. Eventually a combination of property taxes, state and federal grants, bridge tolls, and a 0.5 percent sales tax in the three BART counties provided for completion of the $1.6 billion project (McDonald and Smart 1975).

It was expected that fare-box revenues would both provide for operation expenses and cover the debt for the train cars purchased. This belief was held in spite of evidence presented to the BART board of directors that not a single transit system in North America was operating without a subsidy (McDonald and Smart 1975). In fact, operating expenses the first year after the full system was in service were six times the fare revenues (Webber 1976). State money had to be provided to bail the system out, and the number of rail cars purchased was reduced. This placed no operating constraint on the system, however, since the $150 million computerized train-control system could not handle even the reduced number of cars that were purchased if they were all operating at the same time. The train-control system did not work because of rust on the rails, which impaired detection of the cars, which forced BART to space them farther apart, which reduced capacity. Furthermore, capacity was much lower than predicted because braking rates were calculated without the possibility of rain on the rails. Also, there have been brake-failure problems (BART 1980). BART has operated with ridership far below projected levels of even its physical capacity. The subsidy currently required to keep the system running is about $50 million each year, which covers the portion of operating expenses ($78 million) exceeding fare-box revenues (1978 figures) (Goode 1980, 1981). A regressive one-half percent sales tax in the three counties pays for 80 percent of the annual operating subsidy.

Efforts to Increase Land-Use Densities

The Association of Bay Area Governments (ABAG) once handled all transportation planning and coordination for the region. The bitter dispute that erupted between AC (Alameda-Contra Costa) Transit, which had been responsible for providing most transit service in the East Bay, and BART was not capable of being resolved within the bounds of a voluntary organization like ABAG. The Metropolitan Transportation Commission (MTC) was then formed by the state legislature to study trip needs in the region, handle planning, and serve as a conduit for funds.

ABAG's involvement is currently limited to encouraging local planning agencies to support transit through zoning, at times attempting to bring about density increases through control of the mandated review process for federally funded sewer systems. Even this control has had limited success since at least one Bay Area community promised to upzone, built the sewer system with the funds channeled through ABAG, and then ignored its obligation to increase densities. When challenged on this point by ABAG staff members, that city's response could be characterized as, "You're right. So what?" (Anonymous E).

Density increases in support of transit have long been recognized as a valid goal by ABAG and MTC staff members and is in fact an official policy adopted by the ABAG executive board (Anonymous E). ABAG is a voluntary organization, though, with no direct land-use control in the region's communities. The staff's efforts in urging density increases in Bay Area communities have met with only limited success in some jurisdictions and strong resistance in most.

The BART Office of Planning and Research has long been aware of the benefits more-dense development could have for the system, especially in promoting countercommuting. Little official effort by the BART board or staff has been directed at this issue though.

BART is currently seeking funding for a new $250 million computerized train-control system that will allow tighter spacing of trains in the CBDs of San Francisco and Oakland so the system can carry some of the expected additional riders the continuing office construction boom in those two cities will generate.

Land Use and Ridership

Methods

First, we performed a field survey of the areas within a one-half mile radius of each of the thirty-four BART stations. Pushkarev and Zupan (1977) establish one-half mile as the maximum radius from which a transit system can hope to draw riders not using feeder bus lines or private automobiles. By their standards, the land-use patterns found around all but eleven of the stations are not dense enough to support a rail rapid-transit system like BART. The fact that ridership levels since the system began operations have been so far below projections supports this observation.

Interviews were then conducted to see what efforts local jurisdictions were making in permitting higher-density development to support BART. Few cities were found to be promoting denser development in a meaningful way in more than a token few acres of the study areas. We then focused on suburban stations because of the opportunity for significant land-use intensification and countercommuting there.

The intent of this part of the research was to correlate residential and commercial land-use changes in the areas near the suburban BART stations with ridership changes at those stations. We sought to determine whether commercial development would have an effect on the number of workers riding BART to the nearby station in the morning and whether residential development would be reflected in an increase in riders returning home from work on the evening trains. We chose the Concord line because of the wide range of land intensities near the stations. The outer five stations, beyond the Oakland urban area, were examined.

Data were collected from the BART office of Planning and Research on the number of riders exiting each station on a "typical ridership day" in each quarter in 1976, the year the BART system was completed with the opening of the transbay tube (the Concord line was opened in 1972). These figures were averaged to give yearly mean ridership for the 6–9 A.M. morning rush, the 4–7 P.M. evening rush, and total ridership. Actual means were not compiled by BART. They identify typical days each quarter and then discard their raw data. Data from typical days in the first two quarters of 1982 were also compiled and averaged in the same way. Percentage changes (all positive) in ridership for each rush period and the total day were calculated for the period 1976–1982 for each station and for the five-station average.

Land-use data were collected in several ways. The Orinda and Pleasant Hill stations are in Contra Costa County. Planners have searched files for changes in commercial footage and residential units. We conducted exhaustive searches of the building-permit records in the cities of Lafayette and Walnut Creek. The results were verified by planners in each city. A senior planner in Concord reported that land-use changes had been negligible around that station during the last ten years.

Finally, a political case study was conducted in one station area where particularly strong resistance to land-use intensification was evident. This information was gathered to permit us to make regional planning recommendations with some knowledge of local planning motives.

Survey of Land Uses near Stations

For purposes of categorization, the thirty-four BART stations are grouped here according to intensity of land use within a one-half mile radius of the stations and proximity to city centers. Three urban and three suburban categories are used.

Category	Stations	Characteristics
A	Embarcadero Montgomery Street Powell Street Civic Center Nineteenth Street, Oakland Twelfth Street, Oakland Shattuck	Urban center stations with very high levels of commercial-retail-office use often displacing residential use entirely; residential use medium to high density if present at all
B	Sixteenth Street, Mission Twenty-fourth Street, Mission Lake Merritt MacArthur	Urban high-density residential stations are in areas with two- to five-story dwellings, often split into flats and apartments
C	Glen Park Balboa Park Daly City West Oakland Fruitvale Coliseum Rockridge Ashby	Urban medium density areas with development complete usually before World War II; single-family dwellings on small lots, often built with common walls
D	San Leandro Hayward Lafayette Walnut Creek Richmond	Suburban center stations are near the existing downtown areas of suburban cities; low- or medium-density residential use is common, often limited by commercial activity.
E	North Berkeley El Cerrito Plaza El Cerrito Del Norte Concord	Suburban low-density stations are in areas usually completely developed when BART was constructed, with single-family homes
F	Bayfair South Hayward Union City Fremont Orinda Pleasant Hill	Suburban low-density/rural stations differ from category E stations by virtue of large parcels of open land nearby at the time BART was constructed. Residential use existing then was primarily low-density single-family homes.

The rural stations are not the only ones where development potential still exists. Many of the other suburban stations have what would appear to be prime parcels, larger than an acre and located within a few hundred feet of the BART station, that are vacant. Low demand for the allowed development, lack of proper zoning (often caused by resident opposition to medium- and high-density uses), and speculation-fueled prices that are unreasonably high have all prevented development. This lack of suburban development is having a negative effect on the direction of travel predominant among BART's passengers and on the subsidy required to keep the system running.

Overview of Land-Use Politics near the Stations

Many of the thirteen cities and two counties with land-use jurisdiction in the areas around the BART stations have policies supporting denser development near the stations. In most of these localities, the policies have not moved off the pages of the general plan in a meaningful way. Usually only a small token area is actually zoned for high-density development. Moreover, this often means only about twenty units per acre, as in Contra Costa County, which has land-use authority in the areas around the Pleasant Hill and Orinda stations.

The problem is not policy formulation; it is implementation. Most of the local jurisdictions do not take their land-use intensification policies seriously at all, giving them lip-service and a few token areas of townhouse zoning. No concerted push to promote truly high-density mixed-use or residential development near the BART stations and to discourage development in outlying areas is evident. Interviews with planning staff persons in all jurisdictions revealed that although many of the BART-served jurisdictions have policies promoting supportive development around their stations, efforts to implement these policies have often had limited success because of community resistance. Oakland, for instance, supports such high-density use, but a Rockridge home owners' group fought such zoning and kept the existing neighborhood intact. Concord city planners were forced to include an existing single-family area adjacent to the BART stations in a neighborhood preservation area to quell residents' fears about development. The entire city of Berkeley was downzoned to reflect existing development levels after residents banded together to fight higher-density use at two of the city's stations. Residents in the Mission District in San Francisco defeated a proposed BART-supported redevelopment project in their neighborhood, brought an end to high-density zoning near their stations, and sparked a citywide downzoning similar to Berkeley's. Resident protests in Daly City killed a proposed high-rise project adjacent to the BART station that was supported by city policy and brought about a downzoning in the area from thirty to eighteen units per acre. El Cerrito city planners' attempts to enlarge the medium-density zoning area around the BART stations were defeated because of strong home-owner resistance in those existing single-family neighborhoods. Home owners' organizations in Orinda and Pleasant Hill have prevented the Contra Costa County planners from implementing the county general plan policy supporting high-density development around those BART stations.

Dingemans (1978) has shown that townhouse development in central Contra Costa County (along the Concord line) from 1962 through 1976 (14,229 units) did not concentrate near the BART stations in Concord, Pleasant Hill, Walnut Creek, Lafayette, and Orinda. The average distance of projects from the nearest station was 4.7 miles. Only 10 percent of Concord station riders in 1976 walked to the station. Dingemans generally discussed homeowners groups' opposition to medium-density development that occurred in all of these communities, especially during the 1970s. The two large shopping centers built during this period were two miles from the Concord station. Office construction in Concord and Walnut Creek has generally occurred about a mile from the stations. Problems inhibiting development near these five BART stations were found to be the lack of large vacant parcels and the preemption of land for park-and-ride parking lots. He concluded that the ease of auto access throughout the region eliminates any strong locational advantage for areas near stations.

Land Use-Intensification and Ridership Increases

We quantified land-use intensification and ridership changes at the five suburban stations on the Concord line: Orinda, Lafayette, Walnut Creek, Pleasant Hill, and Concord. Land-use changes were tabulated from January 1972, shortly before the line opened, to July 1982. Ridership changes were calculated from January 1976, shortly after the transbay tube to San Francisco opened and ridership became comparable to subsequent years, to July 1982.

Our hypothesis that more-intensive land use will bring about greater changes in ridership does not hold up. It is not even possible to support an argument that the commercial development that has taken place around these suburban stations, totaling roughly 1 million square feet of work space in ten years, is promoting any significant countercommuting (see table 5-1).

Considering morning (journey-to-work) exits, we would expect the town with the most commercial development to have the greatest increases in ridership. In the case of Walnut Creek, this is true. With 497,910 square feet of work space added, morning exits went up 156 percent. But Pleasant Hill and Concord, with little or no development, had morning exit increases of 130 and 123 percent, respectively. Lafayette and Orinda, with 263,670 and 111,000 square feet of commercial development, had much lower morning exit in-

Table 5-1
Land Use and Ridership Changes, Concord Line

| | Ridership (Exits) | | | | | | Percent Change | | | Land-Use Changes (1972–1982) | |
| | 1976 | | | 1982 | | | | | | Square Feet | Units |
	Morning	Evening	Total	Morning	Evening	Total	Morning	Evening	Total	Commercial	Residential
Orinda	145	1,058	1,770	231	1,371	2,329	59	29	32	111,000	150
Lafayette	113	1,335	2,162	192	1,611	2,641	70	21	22	263,670	125
Walnut Creek	176	2,063	3,566	452	2,749	5,017	156	33	40	497,910	75
Pleasant Hill	108	2,194	3,244	249	3,419	5,118	130	56	58	76,000	704
Concord	155	2,736	4,408	346	4,445	7,025	123	62	59	0	0
Total	698	9,396	15,145	1,469	13,593	22,129	110	45	46	—	—

creases of 70 and 59 percent, far below the five-station average of 110 percent.

Evening exits also do not correlate strongly with increases in residential units. Concord, with no residential units added, has the highest evening exit increase, at 62 percent. Pleasant Hill, with the most additional housing units at 704, has the next highest ridership increase: 56 percent. But Orinda and Lafayette, with 150 and 125 new residential units, respectively, have the lowest changes in evening exits of 29 and 21 percent.

A look inside the station-by-station (entry-exit) matrix provided by BART is even more illuminating, showing that commercial development away from the CBDs of Oakland and San Francisco does not promote countercommuting to work. For example, the majority of riders exiting in the morning at the Walnut Creek and Lafayette stations, which together have seen over three-quarters of a million square feet of new commercial development near them, are getting off inbound trains after boarding at Concord, the end of the line.

Figures 5-1 and 5-2 show the low correlations between land-use intensification and ridership increases. This preliminary bivariate analysis shows that factors other than land-use intensification near stations are primarily determining ridership. These more-important variables include connecting

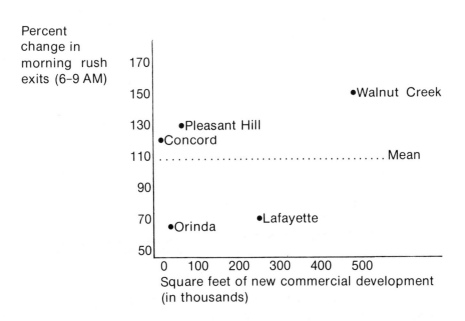

Figure 5-1. Relationship between Land-Use Intensification and Ridership: Morning Rush Hour

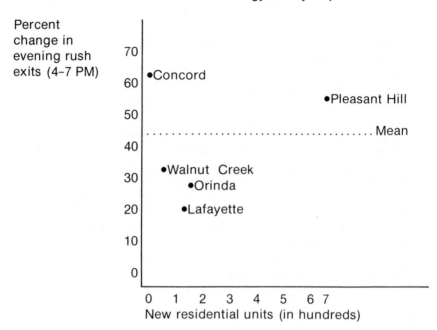

Figure 5-2. Relationship between Land-Use Intensification and Ridership: Evening Rush Hour

bus ridership, freeway access, parking availability, and development in outlying areas. Our data indicate that BART riders continue to locate their residences outward from their jobs, on relatively less-expensive lands efficiently accessible by bus and auto.

As the freeways that parallel all of the BART lines become more congested in the future, we would expect significant ridership increases on BART. As BART reaches its inward-commuting train-car capacity (which it has nearly done) and morning waiting periods at outlying stations increase, we would expect people to avoid taking jobs in San Francisco or to hold out for higher pay so that they can afford the higher rent of closer-in residential areas.

A Supportive Zoning Scenario and Its Consequences for
BART Revenues

We projected transit-supportive zoning for the eleven suburban stations in areas expecting population growth in the next ten years. Table 5-2 summarizes the data used to predict the impact of these future land-use changes on BART revenues. The population figures for 1980 and 1990 come from an ABAG study (1980) detailing anticipated growth in every community in the Bay Area between 1975 and 2000. The antigrowth policies of some of the communities (like Orinda or Lafayette) obviously affect the size of the expected increases because the ABAG planners making the predictions were aware of those policies. This has the effect of reducing the apparent benefits of intensification, as calculated here, since total expected growth in these areas is so low. For the sake of consistency, though, the calculations here are made on the assumption that these cities, like all others, could take steps to direct one-half of the growth they wish to allow within their jurisdictions to the year 1990 to the BART station areas (within a one-half-mile radius). As indicated in table 5–2, the calculations for growth around the two Hayward stations were made differently. In this case, it was arbitrarily assumed that the city's increment of growth would be directed one-third to each station and the final third elsewhere.

The figures for the percentage of land in the area surrounding each station to be devoted to residential uses are the result of a crude field survey. The 50 percent value reflects the existing norm for these low-density suburban areas. The lower 30 percent figure indicates the presence in the study area of a significant amount of nonresidential use, as in downtown Walnut Creek or Hayward.

The density estimates are also an approximation. The station areas are split into two basic density categories. An indicated density of ten dwelling units per acre is an attempt to average the low-density land uses in the outlying portions of some of the study areas (densities in Orinda and Lafayette typically are less than one unit per acre) with the medium-density townhouse or apartment construction that has occurred near the stations. Those areas with estimated densities of fifteen units per acre are in existing community business centers, where most uses are of a higher density. The estimates for Fremont and South Hayward reflect a reduction of one-third from the densities typical on the land developed for residential use in those areas. This reduction was made because much of the residentially zoned land in these areas is not yet developed, and we felt that the density average should be calculated with the entire available acreage in mind.

The estimates of the existing populations in the study areas are the result of calculations made at the end of a chain of assumptions about residential acreage and density in the study areas and should be considered tentative until better data are available. Neither the Metropolitan Transportation Commission (MTC), 440 traffic watershed zones, nor the U.S. Census tracts coincide with our study areas around each station. Population is determined by multiplying 2.6 persons per dwelling (ABAG 1980) times our estimated dwellings per acre times 502 acres (in our half-mile radius areas) times percentage residential (see table 5–2).

Table 5-2

Calculations for Estimating the Effects of a Supportive Zoning Scenario on BART Revenues

	Population (thousands)			Percent Acres Residential	Est. Density per acre	Present Population Estimate
				Study Area		
Community	1980	1990	Change			
Concord	110.1	114.6	4.5	50	10	6,531
Daly City	81.0	84.5	3.5	50	10	6,531
Fremont	135.8	157.5	21.7	30	10[b]	3,915
Hayward[a]	119.4	136.9	17.5	30	15	5,878
South Hayward				50	6.6[b]	4,354
Lafayette	26.0	26.6	0.6	50	10	6,531
Orinda	16.6	18.5	1.9	50	10	6,531
Pleasant Hill	31.1	35.1	4.0	50	10	6,531
Richmond	82.6	84.6	2.0	30	15	5,878
Union City	36.9	45.1	8.2	50	10	6,531
Walnut Creek	70.7	81.5	10.8	30	15	5,878
Daily total						

Annual total = $|28,338 \times 291 = \$8.25 \times 10^6$

[a]Growth split one-third each to two station areas, one-third elsewhere.

[b]Cut to two-thirds density because of excess open land in area

Once the current population of the study areas is estimated, it is easy to calculate the percentage increase that the overall projected population increase will bring if local governments zone to direct 50 percent or, in the case of Hayward, 33 percent of their new growth to the BART station areas.

Although many variables such as automobile ownership, size of the CBD, income, and service area have an effect on transit usage, Pushkarev and Zupan (1977) demonstrate a 1.2 to 1 relationship between transit ridership increases and density increases (within a half-mile of train stations), all other things being held equal, in the range of densities being studied here.

Data supplied by the BART Office of Planning and Research provide the number of riders exiting the system at each station on a typical day in 1980. The number of riders hypothetically using the system in 1990 under this model will be current riders times percentage rider increase plus current riders.

BART also provided data for the average fare received from each passenger embarking at each station. By multiplying the appropriate fare times the number of new riders, the increase in revenues the new growth will bring to each station on a typical 1990 weekday (in constant 1980 dollars) can be calculated.

Totaling the expected revenue increases for all eleven stations in this study gives the total revenue increase to the BART system each day from a serious land-use intensification program at these stations alone. Yearly revenues are

Study Area		Station Total Ridership				
Percent Population Change (1980-1990)	Percent Rider Change	1980	1990	Change	Average Fare (1980)	Added Fares (1980)
21	25	5,632	7,040	1,417	$1.467	$ 2,079
16	20	8,296	9,955	1,659	.947	1,571
171	206	4,537	13,883	9,346	1.315	12,290
61	74	4,127	7,181	3,054	1.035	3,161
82	99	2,562	5,098	2,536	1.169	2,965
3	3	1,897	1,954	57	1.150	66
8	10	2,219	2,441	222	1.470	326
19	23	4,093	5,034	941	1.451	1,365
10	13	1,960	2,215	255	.912	233
38	46	3,113	4,545	1,432	1.280	1,833
57	68	3,960	6,653	2,693	1.355	3,649
						$28,338

characteristically 291 times those of a typical weekday (Goode 1980, 1981).

About $8.25 million per year (1980 dollars) conceivably could be added to BART's revenues if suburban governments made a substantial densification effort for ten years in support of the ABAG regional policy. Our scenario certainly represents the upward range of what is feasible. Physically, the population changes are possible, although the assumptions for Fremont are perhaps unrealistic, resulting in an additional population density of over twenty units per acre. Politically these changes are rather farfetched. Not only would there be strong neighborhood resistance to the intensive developments near the stations, but these jurisdictions would have to reduce development on their fringe lands through downzoning and permit phasing, which would be very unpopular.

What portion of this $8.25 million would be required to pay the added cost of hauling the new passengers is uncertain and would depend on a number of variables. The BART system currently has excess capacity, except during the peak rush hours, and even then the cars are full in only one direction. If ABAG and MTC policies designed to promote off-peak and countercommuting are successful, huge increases in ridership and revenues are possible with very slight increases in operating costs, since the marginal cost of carrying one extra passenger on an underutilized train is negligible.

Off-peak commuting can be facilitated by businesses adopting nonstandard working hours, like 10 A.M. to 6 P.M., so that employees will not be using transit or private automobiles during the peak commuting hours. Where possible, the use of flexible scheduling (a program in which employees can vary work arrival and departure times, within limits, as long as the set number of hours are worked each day or week) should help to spread the peak demand for transit.

Countercommuting, or cross-commuting are the terms used to desribe flow in the opposite direction to the main commuter flow, but at the same time. Promoting industrial or commercial activity near suburban residential area stations can result in riders using the system both ways, morning and evening. Like off-peak commuting, this brings increases in total revenues at virtually no extra cost since the trains currently are running nearly empty in the counterflow direction during rush hours. Substantial density increases for the next ten or twenty years at the outlying stations could significantly improve BART's efficiency.

Countercommuting on BART will be slow to develop, however, since (as Webber has pointed out) the BART lines were laid out parallel with existing freeway routes. Persons desiring to countercommute can reach their destination more rapidly by auto since the freeways are not congested in that direction. Perhaps BART could attract countercommute riders by charging them less. Cross-commuting would be difficult to promote on BART because of its radial design. All lines run through Oakland and San Francisco, so part of any cross-commute journey would be with regular commuter flows. Oakland connects the three Eastbay lines (Fremont, Concord, and Richmond) and has a higher potential capacity than San Francisco, at the other end of the transbay tube. Some increase in cross-commuting through Oakland could be possible with station expansion.

Comparing the BART layout and related land-use patterns to the Toronto system is instructive. Since the intense development that occurred around the Toronto subway stations has often been held up as a model for what is possible for a system like BART, we should take a closer look at that city. Several important differences in the situation there are apparent.

The Toronto Experience

Most of the Toronto subway system was built under older streetcar lines, one of which had a history as one of the most heavily utilized transit lines in the world. The subway was laid under this existing use pattern. In contrast, most of the BART track was laid down existing freight rail line and freeway rights-of-way, which had experienced little transit usage (MacDonald and Smart 1975).

Emilio Escudero, a transportation analyst with MTC (1980), points out that Toronto was going through a period of unprecedented economic growth, at times exceeding 20 percent per year, as its subway system was being built and the pressure for any developable land was so great that builders took maximum advantage of the high zoning potential around the stations. In comparson, the Bay Area economy has suffered with the national economy as a whole through three distinct slumps since the BART bond issue was passed. The effect of these slumps and the slow recovery periods in between has not been felt equally in all Bay Area cities but has certainly had some effect on regional development.

Another factor that encouraged such high-density development around the Toronto subway stations was the transit authority's participation in redevelopment through the use of condemnation authority. The cost of assembling the necessary large parcels, prohibitively high to the private developer who lacks condemnation authority, were covered by the transit system and recovered by the sale or long-term leasing of the property (Urban Land Institute 1979). Only Richmond, Hayward, Oakland, and San Francisco, of the cities served by BART, have existing redevelopment programs, and they are strongly focused on high-density land uses in areas that will support transit (Lake 1980; Bush 1980; Erhardt 1980; Uckerman 1980).

An argument has been made that the California life-style is suited to the mild climate and takes advantage of the yards that low-density housing provides. The harsher winters of Canada, on the other hand, drive residents indoors for a greater portion of the year, making apartment or condominium life more acceptable.

One obstacle to high-density development that exists in the Bay Area but did not in Toronto is lack of appropriate zoning, which points out the single greatest obstruction to transit-supportive land use in the Bay Area and the rest of America. In Canadian and European cities with effective mass-transit systems, either the agency responsible for regional transportation planning is also responsible for regional-land use planning, or close coordination exists between the two. If transit officials cannot control local zoning (or at least be assured of its consistency and suitability), they cannot be assured of the high-density land uses, in the right pattern, needed to support high levels of transit service. European planners treat transportation as part of land-use planning (Holmes 1974). In Toronto, the metropolitan government handles both functions for the region. In the Bay Area, ABAG and MTC clearly do not.

Land-Use Politics

Local governments generally do not manage land use for regional benefit. The private values of locally influential landowner interests generally dominate (Delafons 1969). "Fiscal zoning" is common where fiscally desirable uses, such as retail shopping and clean forms of employment, are encouraged and fiscally undesirable uses, such as apartments, are discouraged (Babcock 1966;

Downs 1973). Large-lot zoning is employed by many suburban jurisdictions to prohibit entry by low- or moderate-income households.

Frieden (1979), reviewing growth-control ordinances in several Bay Area communities, concluded that citizens in these jurisdictions feared tax increases, school crowding, and loss of scenic views. Frieden (1979) and Brooks (1976) both state that environmental protection is used as an argument to conceal these private reasons. Ellickson (1977) generally agrees, but emphasizes that private home owners engage in growth control in order to increase their property values. Case studies of growth management in four northern California jurisdictions by Johnston (1980) verified these earlier findings but found strong concerns for protecting agricultural enterprises and about traffic congestion.

The Lafayette Case Study of Nonsupportive Zoning

In order to gain a specific understanding of the reasons why local government officials fail to support BART stations with medium- and high-density zoning, we examined Lafayette, where the station is adjacent to large areas of open space, parking lots, and some low-density commercial uses. The city of Lafayette has been incorporated, and therefore responsible for its own land-use controls, for about fifteen years. This case study investigated land-use policy in that city, explored why zoning in support of the ABAG Regional Plan and BART has been resisted, and the effect this has probably had on ridership at the Lafayette station.

History of Land-Use Control in Lafayette: Although Lafayette has been a distinct and established community since the early part of the century, its population was small and its growth rate low until the 1960s (ABAG 1980). As suburban development caught up with the town in that decade, the problems with having zoning and land-use controlled by distant Contra Costa County authorities became apparent to residents. On the basic visual level, the lack of well-conceived, uniform standards for signs, architectural design, setbacks, landscaping, and curbs was turning the central portion of Mount Diablo Boulevard, Lafayette's main street, into a sprawling strip of tacky commercial development (Pfautch 1981). Residential construction was being allowed on visually sensitive ridge slopes and crests and was detracting from the scenic quality of the area (Lafayette General Plan 1973).

In an attempt to participate in the decision-making process at the county level, a local advisory committee was formed in the mid-1960s. This Lafayette Design Project group was formed after the California director of parks spoke to the Lafayette Chamber of Commerce about beautification. He emphasized that local land-use control was the only way a community could guarantee the type of development it wanted (Anonymous F).

The Design Project's first efforts went into fund raising to provide land-scaping and underground utility lines along the central portion of Mount Diablo Boulevard, which was undergoing widening and resurfacing at that time. Next came the construction of a small park on an unused triangle of land adjacent to a major intersection. With these two programs undertaken and completed, the citizens involved had demonstrated to themselves and other residents that local efforts could shape their community. The next step, to gain more control, followed logically (Anonymous F).

Incorporation came in 1968. It was supported by a large portion of the community because of concerns about growth, traffic, the design of the central area, and the impacts the opening of the BART station in central Lafayette might bring. Of interest, because it has affected the type of business development the city encourages, is the fact that the city charter prohibits property taxes being levied to support city services. As one of just a few communities in California with this funding restriction, Lafayette must limit its expenditures to what can be covered by general revenue-sharing funds and the portion of the sales, alcohol, and tobacco taxes collected locally and returned to the city by the state. This fiscal structure makes banks and other service businesses less attractive to the city than those that generate retail sales taxes, and the zoning policies reflect this by limiting service businesses and promoting retail activity (Pfautch 1981; Lafayette General Plan 1973).

Development of the Lafayette General Plan: These downtown policies and those controlling growth, density, and type of development in general are laid out in the Lafayette general plan. Adopted in the fall of 1973, the plan follows the tone of the Lafayette Goals and Policy Committee Report of 1970. While many of the policies in the plan promote denser development, which would help support transit services, they are bound to have limited effect in light of the Goals and Policy Committee's primary goal. That goal is to "preserve and enhance the character of Lafayette as a low density, semi-rural residential community" (Lafayette General Plan 1973). This is the town of Lafayette that an observer sees today, not the one suggested by some of the less-fundamental policies in the general plan. The policies that do seem to support transit are these:

> Lafayette should seek to take advantage of the location of the BART station and the Freeway off-ramps, to maximize the integration of both transportation and development of a well-organized and well-designed Central Area nucleus.
>
> Encourage use of any transportation that reduces pollution.
>
> That in addition to areas presently zoned for multiple dwellings, multiples be permitted as a buffer between commercial, freeway, or public areas and single family zones. This should be accomplished in a well-planned manner to avoid an undesirable hodge-podge type of development.

. . . accommodate a reasonable amount of diversity among citizens of Lafayette in terms of age, income, and cultural background.

That a plan be developed to enable needy residents of Lafayette, such as senior citizens on diminishing fixed incomes, to remain in the community.

[The Central Area shall have] a desirable, functional mix of rental ranges and unit sizes in multi-family developments.

Another arrangement which . . . should be encouraged in appropriate locations in the business district is the provision of dwelling units in multi-story buildings above the stores and other commercial enterprises which occupy the ground floor. [Lafayette 1973]

These policies seem inconsistent with the primary goal of a low-density, semirural village. That conflict is resolved in favor of the large-lot, single-family residence. The zoning map accompanying the general plan or a drive around central Lafayette reveals that only token efforts are being made to implement the seven intensification policies. Only very small amounts of land in the city are zoned for multiple-unit or mixed-use development within a half-mile of the station, while large parcels of vacant land ideally located near the existing commercial area and the BART station are not allowed to receive the appropriate development because of the following policies, which conflict with the intensification policies:

Retain a village character in Lafayette.

Encourage large lots . . . .

Plan all facets of transportation to retain the rural atmosphere.

In planning the ultimate population of Lafayette, vacant areas to be developed for single family dwelling shall be zoned to effect existing density and shall not exceed an average density of two families per acre.

That there be no residential or commercial high-rise structures in Lafayette, high-rise being defined as buildings in excess of three stories, except in a few truly exceptional instances where higher structures or portions of main structures would be necessary for the economic development of the core area and where it would enhance the architectural beauty or setting of the building.

. . . that commercial uses not be allowed in the Deer Hill area. [Lafayette 1973]

These six policies force any efforts to increase densities to conform to the existing low densities and limit the use of some of the best land with a potential for the type of transit-supporting development BART planners dream of. Because the second set of policies are the ones being implemented in Lafayette today, the comment buried in the general plan housing element—"under land use policies of this plan, the impact of BART on residential development is expected to be only moderate"—could not be truer.

Reasons for the Conflicting Policies: In other nearby towns with multiple-unit zoning near the BART stations, proximity to BART has been a factor in home-purchase or rental decisions. A few medium-density apartment complexes near the Pleasant Hill BART station, two stops out the line from Lafayette, were recently converted to condominiums. These units are selling rapidly at moderate cost to mostly young, single or married people unable to afford a large-lot, single-family home who want to use BART to commute to work in San Francisco or Oakland (Anonymous C). This is the housing market the Lafayette general plan claims to be providing for with its policies supporting mixed-use, moderately priced, multiple-unit development. Close examination of zoning maps reveals that the low-density policies are the only ones taken seriously by the city council.

For example, consider that roughly twenty acres of vacant land lie within a few hundred feet of the Lafayette BART station, along Deer Hill Road. Although these parcels are hilly and subject to noise from the nearby freeway, which makes them unlikely candidates for the large-yard, expensive custom homes common in Lafayette, properly designed and screened development here at a density of twenty units per acre could provide housing for roughly 1,000 people. These homes would be within walking distance of the existing shopping and entertainment areas of central Lafayette, as well as BART. The city could accommodate four or five years' population increase at present growth rates, relieve pressure on the sensitive ridges and outlying areas they want to protect, and reduce air pollution and traffic congestion in restricted central Lafayette by dense use of just these few parcels. Such development here would also be consistent with the policy of using multiple-unit housing as a buffer because a large, single-family residential area lies beyond the parcels fronting Deer Hill Road (ABAG 1980; Lafayette 1973). This is where the problem lies.

According to one Lafayette planner, high-density development along Deer Hill Road is "a hot potato, a political issue." At the time the general plan was adopted, a majority of the city council members and three planning commission members lived in the residential area beyond Deer Hill Road. They fought to have development not compatible with their view of what Lafayette should be specifically excluded from their neighborhood (Anonymous A).

In this respect, they are not markedly different from homeowners in other areas of Lafayette who have insisted that the only appropriate use of land in their neighborhoods is large-lot, single-family homes. Even a park, swimming pool, and recreation-center complex was defeated by a neighborhood homeowners' association because it would share land with some apartments and because of the traffic and curfew problems that might be associated with it (Anonymous B).

Lafayette is a small town (population about 25,000), with a lot of contact inevitable between the actors on any side of an issue. Several active homeowners' associations exist in the city. They are "wealthy, articulate, and connected" and put up a formidable front before the planning commission and city council (Anonymous F). In one case, a member of a neighborhood association put up $5,000 for the legal fight against an unwanted project (Anonymous B). People are suddenly vocal when development that threatens the character of their neighborhoods is proposed. Their arguments center on noise, traffic, and quality of life (with property value a thinly disguised reason also), and they have been successful both at electing sympathetic decision makers and arguing effectively for preservation of that cherished "semirural" village (Anonymous D).

Conclusions: The Lafayette general plan on the surface supports denser mixed-use development. The zoning map and lack of serious implementation effort reveal, however, that the intensification policies are given only lip-service. The combined action of the city government and home-owners' groups has served to make preservation of the existing low-density character of the community the primary goal. Despite the general-plan policies supporting intensification around the centrally located BART station, the exclusionary, low-density, slow-growth policies aimed at maintaining the status quo in outlying areas of Lafayette are being implemented in the central area as well.

Lafayette is an extreme case. It is, however, located along a busy freeway and BART corridor and certainly could attract more intensive development to its downtown. Due to the extension of BART lines into low-density suburbs, there are several stations where lands with medium-density zoning lie vacant because of lack of demand, such as in Fremont. Urban blight also prevents development intensification near the West Oakland and Richmond stations.

Conclusions

Future Suburban Residential Development

Several forces are emerging to shift the new-housing market toward smaller units built at higher densities. Real after-tax incomes are dropping in the United States, and common sense, as well as urban economics, tells us that as we get poorer, we will consume less housing. Land and finance costs are rising in real terms. Real energy costs are climbing and will make heating and cooling large homes very expensive. The population is aging, which will decrease demand for single-family detached houses. Finally, some cities are moving to limit sprawl and peripheral growth for a variety of reasons, including preservation of agricultural land, service-cost reduction, and view and parkland preservation.

Given enough time and some degree of recovery in the housing market, the acres of empty land with medium-density zoning near the stations on the Fremont line probably will be developed, as will similar areas near other suburban stations. Speculation, lack of demand, the housing slump, and lots of similarly zoned but cheaper land elsewhere in the region, accessible by auto, are combining to stall development near these stations at this time. Hopefully, the cities involved will not downzone these parcels for low-density, single-family homes, which would make these areas expensive to redevelop, if at a later date more-intensive uses were desired.

Future Downtown San Francisco Development

The CBDs of San Francisco and Oakland continue to grow at a rapid pace. Office space in downtown San Francisco, for example, doubled from 1960 to 1980, going from 35.6 million to 71 million square feet. Another 30 million square feet is projected to be developed in the 1980s, of which over half had been approved by mid-1981. The additional riders BART was projected to get from a single large project, the Yerba Buena Redevelopment Project south of Market Street, would require a second transbay tube line, running at capacity with each train full for two hours each evening just to clear the platforms. Since the magnitude of this growth far outstrips anything that is taking place at the far end of the lines, BART will find itself in the position of upgrading its service levels to accommodate one-way, peak rush-hour riders, while the trains return in the other direction nearly empty and are used at only a fraction of their capacity for the rest of the day.

One should also consider the fact that the BART riders represent just the tip of the commuter iceberg. In round figures, BART carries 5 percent of the transbay traffic, and buses carry another 10 percent. The remaining 85 percent still come by private automobile. So for every person riding BART, nearly twenty will be trying to squeeze onto the highways and streets (Webber 1976).

Conclusions Concerning BART and Related Land Use

We cannot undo the mistake that was made by building a heavy rail transit system through low-density suburbs and urban-fringe rural areas. We can, however, strive to make much better use of the existing, costly system (Dingemans 1978). An areawide effort should be undertaken to increase the allowed (and minimum) densities for residential and mixed-use development near the BART stations and to reduce development elsewhere. This does not mean rebuilding existing, stable neighborhoods. Enough vacant land is available near the stations to handle a significant percentage of the Bay Area's anticipated growth. This is especially true in the outlying areas where countercommute-inducing office and shopping development would be most beneficial.

This effort should include any regional means of effectuation that might work: reducing service at stations until land around them is upzoned, using ABAG staff to educate the local governments about the benefits they can obtain by increasing densities, tying state and federal grants for sewers and highways to transit-supportive zoning, and funding those projects incrementally so continued compliance is assured.

In order to facilitate land assembly and redevelopment, BART should be granted redevelopment authority by the state legislature. Joint development of sites immediately adjacent to stations should be encouraged by BART with favorable lease terms on lands owned by it and currently used for park-and-ride lots. Parking could be handled in structures. No additional freeway improvements should be undertaken if those links compete with BART. Some regional land-use control authority may need to be established in order to stop the granting of high-rise office building permits in San Francisco and to encourage or require zoning for offices in Richmond, Concord, and Fremont. Perhaps the MTC and ABAG need the authority to plan land use and transportation jointly as is done in many of the other developed countries of the world.

Directing a large portion of future development to intensive centers at Concord, Fremont, and Richmond would improve BART loading, and therefore transportation efficiency in the Bay Area, in only ten to twenty years. Energy savings would result within each of these cities in the next few years, however, as more energy-efficient structures are built and as people increasingly move around within the centers by bus, bicycle, moped, and on foot. It could be expected, also, that a larger share of households could live in medium- and high-density apartments and condominiums, well served by trains and buses; these families could function with one auto or even none.

General Conclusions Regarding Land Use and Transit

Theoretical and empirical research strongly indicate that a multicentered urban structure will support energy-efficient structures and the use of buses, bicycles, and mopeds for commuting. Heavy rail should not be attempted, except in cases where the largest CBD contains well over 50 million square feet of nonresidential space (Pushkarev and Zupan 1977). Heavy rail systems should serve the largest CBD well, with a loop pattern, and not be extended beyond large cities and densely developed corridors a few miles away. Lines should be extended in phases, so development is concentrated around a few stations at a time (Urban Land Institute 1979; Knight and Trygg 1975). Rail systems should not compete with freeway radials (Webber 1976).

For most urban areas, bus transit, supported by medium-density zoning along selected corridors, will be more cost effective and much more energy efficient than trains or autos. Light rail loops may be efficient in the CBDs of urban areas of 1 million population or more, if well connected to feeder bus lines. Regional transit districts need to negotiate with the various communities in their regions to reach agreement for supportive zoning patterns along transit corridors. One major advantage of bus systems over light rail is that routes can be altered to penalize local governments that do not provide appropriate zoning along the bus corridors.

As the centrifugal growth process of American cities, caused by the freedom of the automobile, slows down in the next two decades in response to slowly changing transportation and housing preferences, land-use intensities will be increased through infill development and redevelopment. Neighborhood opposition will often be strong, especially in stable, higher-income areas. Urban planners and elected officials need to direct infill growth carefully to selected areas of moderate opposition, near train stops and along major bus corridors. Cities and neighborhoods that refuse to permit intensification of land uses should not be served by regional transit services. Stable patterns of transit service, backed by intensive land-use centers and corridors, will permit an increasing number of households to live in energy-efficient dwellings and exist without an automobile.

Notes

1. At ten stories and higher, though, the energy cost of elevators and other utilities can outweigh savings from reduced wall surfaces. Byrne and Howland 1980, p. 60.

References

Association of Bay Area Government. 1980. *Projections 79.* Berkeley, Calif. January.

Anonymous A. 1980. Lafayette city planner. Interviewed September 9.

Anonymous B. 1980. Mental-health-care professional who lives and works in Lafayette and is active in one of the home owner's associations.

Anonymous C. 1980. A realtor. Interviewed September 9.

Anonymous D. 1981. Local news reporter. Interviewed April 3.

Anonymous E. 1981. Local attorney. Interviewed February 6.

Anonymous F. 1981. Private citizen active in Lafayette city politics and planning commission member from before incorporation to the present.

Babcock, Richard. 1966. *The Zoning Game: Municipal Practices and Policies.* Madison: University of Wisconsin Press.

"BART Spurs South County Realty." *Oakland Tribune,* August 6, 1967.

"BART Area Investment Is Booming." *Oakland Tribune,* March 9, 1969.

Bay Area Rapid Transit District. 1980. "System Performance Study." Oakland, Calif.

Brooks, Mary. 1976. *Housing Equity and Environmental Protection: The Needless Conflict.* Chicago: American Institute of Planners.

Brown, Lester; Flavin, Christopher; and Norman, Colin. 1979. *The Future of the Automobile in an Oil-Short World.* Washington, D.C.: Worldwatch Institute.

Bush, J. 1980. Redevelopment director, City of Hayward. Interviewed September 10.

Byrne, Robert M., and Howland, Libby. 1980. *Background Information Summary.* Washington, D.C.: Urban Land Institute.

Council on Environmental Quality. 1974. *The Costs of Sprawl.* Stock # 4111-00022. Washington, D.C.: Government Printing Office, April.

Craig, Paul P. 1976. "Mass Transportation and Energy Conservation and City Structures: The Need for New Goals." Paper presented at Oakridge Associated Universities Energy Symposium, Oakridge, Tenn., October.

Delafons, John. 1969. *Land-Use Controls in the United States.* Cambridge, Mass.: MIT Press.

Dingemans, Dennis J. 1978. "Rapid Transit and Suburban Residential Land Use." *Traffic Quarterly* (April): 289–306.

Downs, Anthony. 1973. *Opening Up the Suburbs.* New Haven: Yale University Press.

Dunn, James A. 1981. *Miles to Go.* Cambridge, Mass.: MIT Press.

Ellickson, Robert. 1977. "Suburban Growth Controls: An Economic and Legal Analysis." *Yale Law Journal* 86 (January): 385–511.

Erhardt, Frank. 1980. Senior planner, city of Oakland. Interviewed September 10.

Escudero, Emilio. 1980. Transportation analyst, Metropolitan Transportation Commission. Interviewed July 25.

Frieden, Bernard. 1979. *The Environmental Protection Hustle.* Cambridge, Mass.: MIT Press.

Goode, Howard. 1980, 1981. Director, BART Office of Planning and Research; member of Rockridge Homeowner's Association. Interviewed August 8, 1980, February 2, 1981.

Hannon, Bruce. 1977. "Reply to Lave." *Science* 195 (February): 595.

Holmes, Edward G. 1974. *Coordination of Urban Development.* Prepared for Department of Transportation. Stock # 5001-00076. Washington, D.C.: Government Printing Office.

Johnston, Robert A. 1980. "The Politics of Local Growth Control." *Policy Studies Journal* 9 (Winter): 427–439.

Keyes, Dale L., and Peterson, George E. 1980. *Urban Development Patterns.* The Urban Institute.

Keyes, Dale L. 1976. *Energy and Land Use: An Instrument of U.S. Conservation Policy?* Energy Policy.

Knight, Robert L., and Trygg, Lisa L. 1975. *Land Use Impacts of Rapid Transit: Implications of Recent Experience.* Document # DOT-OS-30176. Washington, D.C.: Government Printing Office.

Lafayette General Plan. 1977.

Lake, Richard. 1980. Planner for city of Richmond. Interviewed September 11.

Lave, Charles A. 1977. "Negative Impact of Modern Rail Transit Systems." *Science* 195 (February): 595.

Lovins, Amory B., and Lovins, Hunter L. 1980. *Energy/War: Breaking the Nuclear Link.* San Francisco: Friends of the Earth.

McDonald, Larry L. 1980. Planner for city of Daly. Interviewed September 10.

McDonald and Smart, Inc. 1975. *Key Decisions: A History of the Development of Bay Area Rapid Transit.* Final Report to Department of Transportation. Report DOT-OS-30176. Washington, D.C.: Government Printing Office.

Metropolitan Transportation Commission. 1978. *Land Use and Urban Development Impacts of BART.* Final Report to Department of Transportation. Report DOT-BIP-FR-14-5-78. Washington, D.C.: Government Printing Office.

———. 1979. *The Local Implications of BART Development.* Final Report to Department of Transportation. Washington D.C.: Government Printing Office.

Peskin, Robert L., and Schofer, Joseph L. 1977. "The Impacts of Urban Transportation and Land Use Policies on Transportation Energy Consumption." Report DOT-OS-50118. U.S. Department of Transportation, April.

Pfautch, Lee. 1981. Member of Lafayette's Planning Commission. Interviewed on April 3.

Pushkarev, Boris S., and Zupan, Jeffrey M. 1977. *Public Transportation and Land Use Policy.* Bloomington: Indiana University Press.

Roberts, Jerry. "How S.F. High Rises Sprouted—65 in the Last 10 Years." *San Francisco Chronicle,* May 12, 1981, p. 3.

Stobaugh, Robert, and Yergin, Daniel, eds. 1979. *Energy Future: Report of the Energy Project at the Harvard Business School.* New York: Random Books.

Uckerman, Franz. 1980. Planner for the city of San Francisco. Interviewed September 10.

U.S. General Accounting Office. 1981. "Greater Efficiency Can Be Achieved through Land Use Management." Report EMD-82-1. Gaithersburg, Md.

_____. Solar Energy Research Institute. 1981. "Building a Sustainable Energy Future." Committee Print, House Committee on Energy and Commerce. Washington, D.C.: Government Printing Office.

Webber, Melvin. 1976. *The BART Experience—What Have We Learned?* Monograph 26. Berkeley: Institute of Transportation Studies, University of California, Berkeley.

6

Small-Scale Technology and the Electric Power Industry: Law, Politics, and Economics in State and Federal Policymaking

Michael F. Sheehan

This chapter analyzes the response of the electric utility industry to the changing needs of American communities in a period of substantial economic stress. The critical issue to be investigated is whether the reluctance of both regulators and the industry to adapt themselves to new conditions is based on some legitimate view of the public interest or is rooted instead in a desire to protect the pecuniary and status interests of the entities involved regardless of the impact on the general welfare.

Early Development of the Electric Power Industry

From the beginning, control of the electric utility function has followed two divergent paths: corporate control under publicly granted franchise or regulation and municipal ownership. Of the two, private ownership has always generated the bulk of the power and supplied the majority of customers. When small generating systems were the rule, regulation at the municipal level was standard. At first this took the form of simple contracts to provide for municipal lighting purposes and allow use of streets and alleys for power lines. During the reforms of the progressive era, local regulation was strengthened to include municipal regulation of rates charged to private customers as well. With the growth of multijurisdiction utility companies, movement toward state regulation began in 1907[1], which rapidly became the dominant regulatory overlay[2] with respect to retail rates.

In this period of rapid development, several forces were at work. The provision of electric power had passed out of the stage of being a luxury good or novelty and was rapidly becoming a residential, commercial, and industrial necessity. As such, its status as a business "affected with the public interest" made imperative the need for public input, if not control, in securing adequate service at equitable rates. Other factors at work strengthened this trend. Electric utilities required the use of public streets for poles and lines and could make a powerful case that it was inefficient to have multiple sets of poles and lines of competing companies cluttering the city thoroughfares. The acceptance of this argument led to the granting—in effect, if not in law—of geographic monopolies, which further strengthened the case for regulation.

In terms of the economics of electric generation, the industry was on the verge of discovering how to reduce costs substantially through corporate consolidation, load diversification, and the construction of larger, more-cost-effective generating stations to serve expanded markets. Pressures at the local level for nonduplication of lines and at the regional level for corporate consolidation and market expansion led inexorably to the creation of powerful geographic monopolies. Yet these monopolies were not created by the vested interests controlling them solely in the interest of providing good service at rates reasonably reflecting costs. With costs falling rapidly, the ability to keep rates from falling as rapidly held the potential for the taking of vast profits. Since the regulation of rates depends fundamentally on the ability of regulators to determine the costs of providing service, a powerful incentive exists to create a corporate structure large enough to field a team of expert attorneys, accountants, and other professionals superior to the regulatory staffs of all but the largest cities, and jurisdictionally diverse and organizationally complex enough to make it virtually impossible for regulators to determine the real costs of providing service within the local jurisdiction.

The movement toward state-level regulation was a direct response to this organizational offensive on the part of the utilities.[3] Yet regulation even at the state level was soon overwhelmed by the consolidation of large state systems into holding companies and the pyramiding of first-level holding companies into second-, third-, and higher-level edifices under the leadership of men like Samuel Insull.[4]

At this point the organizational consolidation of the industry surpassed the bounds of any economic or engineering justification. Higher-level holding companies were purely paper organizations whose assets were simply the stock of the next-lower-level holding company. The holding-company structure was designed to accommodate a variety of specific financial and regulatory abuses. The largest cost to an operating company is the cost of capital, seen in one sense as the actual value of its physical capital in use and in another as the purchase price less depreciation of its assets, or alternatively, as the combined value of the operating company's outstanding stocks and bonds.

Since the structure of the holding companies tended to be built from the operating company upward, this allowed holding-company stock to be sold to the public based on the holding company's purchase of operating stock at prices sharply in excess of the actual supporting book values. This was functional in two ways. Since the price the public was willing to pay for holding-company stock depended upon the value of the assets of the holding company —that is, the stock of the operating company—a high intercorporate transfer price, which could be touted as a valid indication of real market value, would facilitate the sale of holding-company stock at excessive prices, resulting in a windfall for the organizers at the expense of investors. The second complementary aspect of this policy was that if regulators could be bludgeoned into accepting the sale price of stock as a major basis of valuation, rates would have to be raised to reflect the higher prices in these paper transactions. And even if these prices were not accepted, tremendous pressure was put on the management of the operating companies to pay dividends that reflected the going rate of return applied to the inflated value of the stock. In a period of falling costs per kilowatt hour, this resulted in a strong tendency to disguise cost reductions in order to maintain rates, thus allowing a surplus for higher dividends.[5]

The holding companies also fully utilized the mechanism of "contracts for management services" to skim off profits of the operating companies at ratepayers' expense. In practice, this meant that operating-company managers were simply ordered to contract with the holding company for various sorts of financial or advisory services at excessive prices. These charges were then justified as legitimate costs of doing business and passed on to ratepayers.[6]

The holding companies also exploited the operating companies by forcing them to make loans to either the holding company itself or other subsidiaries at below-market rates. Often the operating company would be forced to borrow the money at the expense of its own credit rating.[7] In addition, many holding companies then required (and many still do)[8] that the regulated operating companies buy equipment and services from unregulated subsidiaries at prices reflective of the less than arm's-length circumstances of the sale.[9]

Finally, the multistate, multiple-level holding company, controlling operating companies in a crazy-quilt geographical pattern within a maze of intercorporate agreements affecting sales, equipment purchases, intercorporate loans, security transactions, and a host of other complications, with each state commission having jurisdiction only over the local operating company, was ideal for creating an atmosphere where effective regulation of prices and profits could be effectively nullified.

It had become clear by the late 1920s and early 1930s that federal regulation of some sort was in order. With the election of Franklin Roosevelt to the presidency in 1932, the stage was set for the passage of federal legislation dealing with some of the major complaints against the holding companies. The Rural Electrification Administration, with financial help from the Works Progress Administration (WPA), acted to provide electric service to farming regions of the country that the utilities had refused to serve, and the Tennessee Valley Authority was created as both a regional development agency and as a yardstick power producer to provide information to regulators on the legitimate costs of producing power.

In 1933 the Securities Act was passed to effect reforms for the protection of stock and bond holders, and in 1935, after a long and bitter fight, the Public

Utility Holding Company Act was passed to break up all second-level or higher utility holding companies (holding companies of holding companies) and effect the reorganization of holding companies to produce sensible patterns of operating-company systems that would embody all available economic, engineering, and operating efficiencies. This forced reorganization of higher-level holding companies, the famous "death sentence" provision of the act, was effective in reducing the organizational conglomeration of the utility industry for illegitimate ends and limited future higher-level organization to the maximum level required for engineering and economic efficiency in the public interest.

Since one of the major regulatory problems of state commissions was their inability to regulate interstate combinations of companies—the scope of business organization had progressed beyond the scope of regulatory control —Congress also acted to expand the authority of the Federal Power Commission to include regulatory control over the sale and transmission of electricity in interstate commerce.[10] Taken together, these several pieces of legislation were effective in correcting the most-glaring abuses in the electric utility industry and for creating new regulatory tools to meet new problems as they arose.

Throughout these struggles on the part of both the regulators and the industry, from the early beginnings until virtually the energy crisis of the 1970s, the utilities constituted a declining-cost industry. As output increased and scale of operation could be expanded, costs fell. The predominant regulatory problem under such an assumption was to see that these cost savings were passed on to consumers.

From the time of the passage of the Holding Company Act (and the creation of the Securities and Exchange Commission to supervise it) and associated legislation until about 1973 and the first energy crisis, the utility industry enjoyed its golden age. Once the battles of the 1930s and early 1940s were fought and won, a long period of relative tranquility ensued among utilities, regulators, and the public, for several reasons. First, large economies of scale based on expanded markets and favorable technological advances resulted in declining costs per unit of electricity, which allowed the companies to post a long series of rate decreases. With consumers quiescent, there was little pressure on regulators to extract all possible cost savings from the companies or to insist on technological innovation in directions away from the central-station orientation of the companies. Also, from approximately 1940 until the rapid inflation around the period of the Vietnam War, capital markets were fairly stable and utility construction programs could be financed at low interest rates. Given the capital intensity of the utility industry, this was no small advantage. Low interest rates and substantial and ongoing economies of scale created a mode of thinking among utility executives and regulators that capital expansion was the key to prosperity; newer, bigger, and better power plants were seen, virtually as a matter of ideology, as the key to the maintenance of prosperity. Smaller plants, decentralization of generation, and conservation were seen as the road to higher average cost (cost per unit of output), restricted output, and sharply higher rates for consumers.

This view, based on a particular set of historical circumstances, was to blind, utility planners in a later era for the need to change their approach to system planning. Regulators who had been educated and spent the bulk of their professional careers in the era of low interest rates and continuing economies of scale were unprepared to confront the disruptions faced by the industry and the nation when these conditions suddenly began to change between 1973 and the present.

1973 to the Passage of PURPA

With the Arab-Israeli War of 1973 and the Arab oil boycott, the electric utility industry entered a new era. The prices of both generating-station fuel oil and natural gas began a sharp rise, which was to increase fuel costs several times over during the next several years. The utility industry, extremely sensitive to any potential curtailment of the market resulting from price increases and desirous of maintaining the good public-relations image it had nurtured since the mid-1930s, responded in two ways. It asked for and received permission from most utility commissions to separate the fuel charge out from the balance of the customer's bill, so that it would be perceived as simply a pass-through item whose rapid and continuing augmentation could be laid at the door of the Arabs and the oil companies. Second, it received permission from the same utility commissions to increase this fuel-adjustment charge without further authorization as utility costs rose. Through this mechanism the usual, if not the only, forum for the articulation of ratepayer grievances was eliminated. The success of these maneuvers is a substantial public-relations victory for the industry.

For a longer term solution to the fuel-oil price problem, utility planners turned, predictably, to large-scale technology. In the years 1973 to 1980, this meant in the main an acceleration of the move toward nuclear power. The years 1973 and 1978 saw the planning and construction of the vast majority of all operating nuclear-generating stations.[11] The utilities anticipated that nuclear-power stations, though expensive to construct, would have such low fuel costs that at high levels of utilization they would be substantially cheaper in terms of costs per kilowatt per hour than fuel-oil or coal plants. Thus it appeared to the utilities, the federal government, and many state regulatory boards that advances in large-scale technology had presented a solution. The industry's period of self-satisfaction was short-lived. With the vast increase in the number of nuclear plants either on line, under construction, or in the active planning stage, domestic uranium prices began to rise sharply, from about $8 in 1976 to about $35 in 1981. In addition, the rise in domestic prices turned the eyes of suppliers toward foreign reserves only to find an international uranium cartel in the making.[12] Because of this it came to be perceived quite rapidly that the turn toward nuclear power of traditional design could not legitimately be based on the slogan of energy independence because we would soon have been reliant on supplies of yellowcake through a foreign cartel, as well as by OPEC in oil.[13]

Once again, a solution seemed to present itself, this time in the form of the breeder reactor. Such reactors have the characteristic of converting standard uranium fuel into plutonium in amounts that exceed the fuel value of the uranium needed to produce it. If the breeder could be made safe, it might indeed have been the solution to the yellowcake price problem. Unfortunately, the specter of virtually unlimited production of such an extremely dangerous product as plutonium, under current conditions, was perceived as being too hazardous to undertake without years of further basic research. The decision to postpone indefinitely the commercial development of breeder technology was made by President Carter early in his administration. The only major vestige of the program of commercial development left today is the Clinch River breeder reactor under construction in Tennessee.

It became clear by the second energy crisis in 1977–1978 that neither OPEC nor increasing fuel prices were likely to relent. Because of the serious effect of high energy prices on all sectors of the economy, Congress was moved to pass the five bills comprising the National Energy Act in 1978.

The Public Utility Regulatory Policies Act (PURPA)

The most important component of the National Energy Act of 1978 was PURPA. Convinced that the pattern of supply and consumption that served our nation best when energy was cheap was not efficient when energy was expensive, Congress passed PURPA to mandate certain reforms at the federal level and to require at least the consideration of certain reforms at the state-commission level.

PURPA was a direct attack on the ideology of demand expansion, on the belief in the economic efficacy of larger and more-expensive generating units, and on the further concentration of utility control of the market. In response to the industry's belief in continuance of economies of scale in generation, PURPA recognized declining load factors, increasing fuel costs, rising costs of new plants, and declining central-station reliability (with the move toward nuclear power.)[14] In response to promotional practices designed to encourage demand—and especially declining block rates—PURPA mandated ratemaking based on the principle of cost of service; each class of customer should be charged rates reflecting the actual cost to the system of providing such service (specifically mentioned as tools for accomplishing this were time-of-day rates, seasonal rates, and peak-load pricing. Finally, in response to the fixation with large-scale technology as the solution to strategic problems, PURPA provided regulatory and financial encouragement for the development of cogeneration and independent small-scale power production.

Congress was explicit in setting PURPA's goals: the encouragement of conservation and maximum efficiency in resource and facility use, and the creation of rate structures that would provide the proper incentives to ratepayers in terms of the timing and magnitude of demand.[15] In devising the wording of the act, it was clear that Congress intended to give generally sleepy state commissions a shove in the direction of reforms along these lines. To help ensure that the required state hearings were not completely dominated by the regulators and the investor-owned utilities, the act stressed the importance of citizen intervenors in the regulatory process and empowered the federal Department of Energy to intervene when necessary to keep the process moving in the proper direction.

Except for the small-scale power provisions, the portion of the act that caused the most controversy related to the reform of rate structure. Under PURPA, declining block rates had to be abolished unless it could be shown that such rates were fair and cost-effective.[16] In order to justify expansion of generating capacity—with its putative lowering of costs per unit due to economies of scale—the utilities had, from the time of Insull, offered rates that favored large users and the expansion of use by all ratepayers. Such a policy could be justified in terms of economic efficiency only in the presence of economies of scale. Absent these scale advantages, promotional rates embody an inherent contradiction in that they encourage higher levels of consumption, which lead necessarily to the need to install more expensive capacity and raise rates. The higher rates lead to demand reductions, resulting in the end in a minimal increase in demand, higher rates, and surplus capacity. Faced with the prospect of reductions in demand due to the prospective need to raise rates, the utilities' response in practice has been to attempt to increase rates to those consumer categories where the subsequent reduction in demand would be smallest (where the demand was most inelastic)—for example, on the customer-charge component of residential rates.

In order to combat this twisting of the rate structure to mitigate the adverse effects of what at root was a problem of technology, Title I of PURPA required that charges to each class of customer reflect the actual costs of serving that class of customer, thus enforcing that basic postulate of regulation that requires that pricing be nondiscriminatory.[17]

Congress indicated that since new generating capacity was turning out to be more expensive than existing capacity (with the possible exception of atomic versus oil-fueled capacity in New England) as a simple matter of the rapidly escalating costs of new base-load power plants, and since it was well known that the provision of power to customers at peak times was substantially more costly than at other times, rates should take these differentials into consideration wherever this could be done cost-effectively. In practice, attention was concentrated on experiments to test the efficacy of seasonal rates (most urban-based utilities face peak demand in the summer due to the air-conditioning load), time-of-day rates, and long-term marginal-cost pricing.

Where Title I of the act was important in forcing commissions and utilities to confront the economic and policy consequences of their actions in terms of pricing, Title II was an effort on the part of Congress to establish the right of other than central-station technologies to participate in providing service to power consumers. Historically, virtually all noncentral-station modes of serving the needs of power consumers had been driven out of the market by the monopoly practices of the large power companies justified on the basis of the availability of large economies of scale. By 1978, however, it was becoming increasingly clear that costs were rising with advances in size instead of falling and that at some point alternatives would once again find a place in the power-supply picture. Based on this, Congress acted to remove the accretion of regulatory constraints on the development of cogeneration and small-scale power production.[18]

Title II is explicit in requiring that utilities purchase power from small-scale producers. This was a revolutionary notion, compounded, from the point of view of the utilities, by language in the act and in the supporting rules published by the Federal Energy Regulatory Commission (FERC) that ordered state commissions to determine what the in-house marginal cost of producing electricity was to the existing utilities and then to set the purchase price of power produced by alternative technologies at that avoided-cost rate.[19] In order to provide an informational basis for determining these rates, Title II and the FERC rules were explicit in requiring utilities to file a great mass of data on generation costs and capital-expansion plans. The utilities considered these to be private if not proprietary, especially after consumer groups began to request access to this information.[20]

In addition to the requirement that state commissions set rates for buy-back power on the basis of avoided costs, Title II also mandated that small-scale power producers be treated fairly in terms of the provision of backup power at reasonable rates and in the requirement that small-scale power producers be allowed mandatory access to the lines of the local utility at reasonable prices in order to wheel output to neighboring utilities offering higher buy-back rates.

The development of small-scale hydroelectric potential was encouraged through several means. Subsidized financing was made available for feasibility studies and for actual conversion, should the feasibility study indicate that the creation of such capacity was cost-effective. Under the preference clause of the Federal Water Power Act of 1920, federal hydropower sites were made available to public entities, and a number of federal agencies assisted with matching grants of various types. For communities that had purchased existing hydroelectric facilities from private power companies wishing to rid themselves of tax and insurance liabilities under the condition that such facilities never again be used to produce electricity, Title II declared that all such restrictions on use were contrary to public policy and void, thereby creating the opportunity of bringing many of these facilities back into production outside the sway of their erstwhile owners.[21]

These provisions, when considered in conjunction with federal encouragement of efforts at conservation and solar-energy development, provided an opportunity for a fundamental reorganization of both sides of the electric demand and supply structure in the United States. Yet these reforms generated strong opposition from powerful vested interests.

The Economics of Electric Power Production since 1978

Organizational Study of the Industry

Much of the current debate revolves, either directly or implicitly, around the issue of whether the major central-station utilities meet the economic and public-policy criteria for being natural monopolies. This question is critical for the determination of whether there is an important role to be played by alternative technologies in the national power picture.

Before discussing what economic theory has determined to be an appropriate analytical definition of natural monopoly, a brief description of the salient characteristics of the industry as it stands is in order. First, the private sector of the industry is, in the main, composed of corporations that engage in three functionally distinct activities: the generation, transmission, and distribution of electricity. In terms of generation, the typical utility operates many generating units of the same type; the average utility will have several base-load plants, several intermediate-load, and usually a number of peaking plants. All utilities in the United States are, in addition, interconnected with grids made up of other generating utilities. This allows them to share capacity, increase reliability, and importantly, participate in the gains of central dispatching of facilities over an entire interconnected system.[22]

In terms of governmental supervision, all private utilities are subject to regulation at the state level for retail sales and at the federal level for wholesale sales. Many utilities whose service areas extend over more than one retail jurisdiction are supervised by more than one state regulatory commission. At the state level (retail), all electric utilities possess official geographical monopolies

over their service territories as a natural concomitant of the legislative determination that they possess the attributes of natural monopoly.

In sum, the relevant organizational characteristics of the typical utility
are: (1) that it is a multiplant firm unable to serve economically the totality of
the demand facing it at existing prices with a single plant; (2) that substantial
capacity sharing exists at the grid, pool, or reliability council level; (3) that
plant dispatching is often done at the grid level in order to take advantage of
the benefits of economy dispatch; and (4) that many retail-level operating
companies are wholly owned by higher-level holding companies that control
one or more of such companies.[23]

Defining Natural Monopoly in a Policy Context

In his popular microeconomics text, C.E. Ferguson explains natural monopoly as lying "in the cost of establishing an efficient production *plant,* especially
in relation to the size of the market. [Natural monopoly] comes into existence
when the minimum average cost of production occurs at a rate of output sufficient, almost sufficient, or more than sufficient to supply the entire market
at a price covering full cost."[24] Thus it is clear that for Ferguson at least, the
relevant unit of analysis is the single plant. Richard Ely, the originator of the
term *natural monopoly,* used a broader test. A natural monopoly exists when

> the returns are in accordance with the law of increasing returns—the greater
> the output the larger the return. The cheaper the rate at which an increasing
> output can be furnished, the greater is the tendency toward monopoly, be-
> cause whenever two competitors unite, they can furnish the service or com-
> modity more cheaply.[25]

More-recent work by economists dealing specifically with public-utility
questions has resulted in a spate of refinements to the concept, all crafted with
an eye toward serving the author's view of the existing situation. James C.
Bonbright, the dean of public-utility economists, in his classic *Principles of
Public Utility Rates,* eschews a cost-based standard in favor of one resting on
the network characteristics of the distribution systems of natural gas and electric utilities:

> What favors a monopoly status for a public utility is not the mere fact that,
> up to a certain point of size, it operates under conditions of decreasing unit
> costs—an attribute of every business, including a farm or a hand laundry.
> Nor is it even due to any indefinite extension of the declining-cost portion of
> a curve relating unit costs of production to scale of output. It is due, rather,
> to the severely localized and hence restricted markets for utility services—
> markets limited because of the necessarily close connection between the util-
> ity plant on the one hand and the consumers' premises on the other.[26]

> What has just been said about the interplay of the factors of economies of scale and of localized markets points to the significance of the fact ... that public utility companies are essentially transmission agencies.[27]

Under this criterion, then, there is no reason to suppose that the generation component of the major utilities is entitled to natural monopoly status even if it were to be operating in a decreasing-cost portion of its long-run average cost curve, especially given the rapid advance in the ability to transmit electric power at very high voltages over (until recently) unheard-of distances. Under the Bonbright formulation, the existence of regional and interregional power grids would imply that single private utilities ought to have multistate monopolies.

Both Schmalensee and Kahn, writing in the 1970s, however, return to a cost-based standard. Schmalensee's rule is that "an industry or activity is said to be a natural monopoly if production is most efficiently done by a single firm or other entity."[28] He elaborates the rule by distinguishing between a temporary natural monopoly, where the single firm is only the least-cost producer at a particular level of production, and a permanent natural monopoly, where a single firm would always be able to produce at the lowest cost.[29] Yet there is an important gloss in Schmalensee's formulation. A casual reading of his notion of a permanent versus a temporary monopoly gives the impression that in the case of a permanent natural monopoly, a particular single firm would maintain the lowest-cost position throughout. Yet it is not at all clear, either theoretically or empirically, that such a situation must or would obtain. The major factor that would interfere would be the unwillingness of the firm to write off vestiges of existing capital that have been made redundant—for example, by a rapid increase in the scale of demand justifying a much larger plant rather than the duplication of an existing size of plant. Since only under competitive pressures or unrealistically strict regulation would the firm be required to write off such capital as an expense to stockholders, as opposed to continuing to include it in the charges to ratepayers, the firm would have ceased to be the lowest-cost producer. In the perfect situation, a new firm could offer the existing level of service with identical plant less the redundant component at a lower cost. Integrating this caveat into Schmalensee's rule results in the following corollary: for a particular firm to be considered a permanent natural monopoly, it must be the lowest-cost producer at every effective level of demand relative to any potential competitor with any combination of capital, at any time. In addition, to qualify for permanence, a firm would not only have to be the lowest-cost producer relative to any combination of smaller firms but also with regard to any larger firm. So, for example, if Iowa is currently served by seven large private utilities, none of them would qualify as a natural monopoly if a single firm could serve the entire state at a cost of production lower than that achieved by any of the seven. Since the nation is dotted with utilities, all enjoying implicitly permanent official monopoly status (presumably based on the fact that they are permanent natural monopolies)—that is, it is being assumed they are the lowest-cost producers at any level of production—then either they all enjoy exactly equal costs of production (which is obviously not the case given their unique and widely varied rate structures) or some large amount of combination ought to be possible that would result in even lower costs to ratepayers. Since in the main, existing service districts and levels of demand facing utilities are the result of historical accident, with the addition of relatively inconsequential amounts of rationalization based on systemwide cost considerations, it can be surmised that very few, if any, existing utilities are at anything approaching optimal size. In fact, virtually no effort has been made to determine what optimal size is in a static sense, given existing conditions, nor could any determination be made about optimal size for the future, given the large amounts of risk and uncertainty prevailing in the market today.

Even a superficial perusal of the electric utility industry in the United States today, with its large amounts of excess or economically obsolete capacity (New England's fuel-oil-fired plants, for example), large sunk costs in cancelled plants, combined with rapidly fluctuating demand, input prices, and relative market prices of competing energy sources, should make it clear that no existing firm could support a claim to being a permanent natural monopoly under the modified Schmalensee rule.

It is now possible to approach the practical question of whether the existing organization of production—the operating companies themselves, plus the grid organization, and the holding companies, combined with the supervision of the regulatory regime—approximates least-cost production. Put another way, is there any combination of facilities or organizational form (limiting consideration for the moment to technologies based on central-station generation) that could serve demand at lower cost?

It has been demonstrated that a strong case can be made that the existing system is not even close to comprising the least-cost combination of facilities (still again within the central-station context) if only because of the large current amount of excess capacity nationwide, as well as in certain specific regions.[30] And although it has been argued that this high level of redundant capacity is in the main due to the presence of large amounts of high-cost fuel-oil capacity in the Northeast—an argument beside the point for current purposes[31]—this ignores the large amounts of excess capacity in the Midwest and elsewhere where there is very little, if any, fuel-oil-fired-based load capacity.[32]

In all of the areas where significant excess capacity exists, it exists in fact because a price that covers costs intersects the demand curve at a quantity that leaves a large fraction of existing capacity idle. Put differently, the earning capacity of existing capital is insufficient to support its valuation at original cost; or yet more bluntly, the excess facilities are not worth what the utility company paid for them. The efforts on the part of utility managements and sympathetic regulators to have the public pay the difference between actual value and original cost is one component of the excess of the current cost of service over the minimum necessary to provide the specified level of service.

Economies of Scale, Rising Capital Costs, and Optimal Planning

It might be argued that a significant amount of the nonfuel-oil-fired excess capacity now extant in the industry was built in anticipation of demand increases in order to gain the benefits in terms of lower-cost service of substantial economies of scale. The cost of idle capacity, so the argument goes, will be outweighed by the savings in terms of the reduced current and future costs of the new facility relative to existing capacity. The argument has several flaws.

In an interconnected system with properly coordinated grid-level capacity planning, the desire to build large and ahead-of-jurisdictional demand in order to obtain economies of scale would be handled through a phased construction program where one utility would build a large plant and share the excess with other utilities, which would thus be able to delay construction of their own new plants. With proper planning, it is difficult to imagine why there would ever be more than the equivalent of one full-size base-load plant worth of excess capacity. Any more than this is a sure indication of poor planning on the part of grid members.

In dissecting the arguments attempting to justify large amounts of excess capacity on the basis of economies of scale, it is important to understand that what is being traded off are the holding costs of the facility, some current and some in the future. In times of low interest rates and large economies of scale, such an argument is especially convincing. Yet in the presence of very high interest rates and capital costs generally, the magnitude of the economies of scale necessary to offset the holding costs—especially since future benefits have to be discounted at the higher effective interest rates—has to be much larger. Rules of thumb based on interest rates and rates of demand growth dating from the era before the energy crises are not likely to be adequate guides to capacity planning under today's conditions and must be recomputed.

This upsetting of the balance between economies of scale and interest rates has not only been due to the rapid increase in the level of the latter over the last few years but is also rooted in the decline in the level of available economies of scale. These economies can be divided into two types. The first results from the lowering of fixed costs per unit of output as plant size increases to serve an adequate level of effective demand. In this regard, the size of the typical nuclear family seems to have reached an effective limit at somewhat less than 1,300 MW, while coal plants seem not to extend much beyond the 600 to 800 MW range.

The second type of economy is based on increases in plant efficiency due to the introduction of new technology. In power plants, one important measure of this is the heat rate (the number of Btu's of heat needed to generate a kilowatt-hour of electric power). For a variety of reasons, including safety and environmental protection, these heat rates have been increasing over time, signaling a decrease in the productive efficiency of the plants, and with sharp and continuing increases in fuel costs, economic efficiency has been falling as well.[33]

To compound these problems, plant construction costs since the late 1970s have escalated at a rate in excess of the general rate of inflation as measured by the industrial-producer price index.[34] Capital costs in terms of the return needed to attract investment have been increasing as well, adding substantially to the large general effect of high and spiraling prices of central-station power facing the public.

Consumer costs have been rapidly increasing for so long that the public expects them to continue escalating unless alternative investments in cost reduction are made. This has led to an emphasis on the reduction in the use of electric power and the switch to other means of serving the same needs on the part of both of consumers generally and industry. For the utilities this has led to lower utilization rates for existing capacity and the operation of facilities at less than their optimal levels of output, resulting in a yet higher level of unit costs because more overhead must be crowded onto fewer and fewer units of power sold.

Falling Load Factors and Falling Reliability Levels

Load factors[35] for the industry at large have been falling over the last several years, indicating a decline in another measure of the efficiency of use of existing capacity.[36] The roots of this deterioration in load factors are to be found in the general constriction of demand, which has affected the average level of demand more heavily than peak demand, and in the general unwillingness of most utilities to adopt peak-load pricing to smooth out peaks. Many utilities have a reason to fear, and a strong incentive to oppose, any policies that would allow a subset of existing installed capacity to serve demand, thereby calling attention to the functional redundancy of the remainder. The problem is highlighted by the fact that excess capacity is computed as that amount of capacity over and above peak load plus 15 percent reserve margin. Any pricing scheme that reduces peak load that had been served by jurisdictional capacity produces excess capacity; on the other hand, any pricing scheme that encourages a heightening of peak demand for a utility suffering from regulatory problems due to excess capacity will tend to mitigate those problems.[37]

In addition to problems with load factors, utilities generally have been suffering from an average decrease in the reliability of base-load capacity as measured by the capacity factor (electrical output divided by plant design capability). In terms of nuclear-power plants, average capacity fell from 66 percent in 1978 to 51 percent in the first half of 1980 for plants above 400 MW, and from 62 to 48 percent over the same years for plants 800 MW or larger.[38] For coal plants, the capacity factor for plants in the 400 to 800 MW range was 62 percent over the period 1968–1977, with smaller plants in the 200 to 400 MW range at 70 percent.[39]

Three points need to be made: plant reliability seems to be falling over time; plant reliability seems to have fallen sharply with increasing size; and since capital costs go on and are charged to consumers (along with the higher costs of replacement power) even when facilities are not available for dispatch, the rates must rise when capacity factors fall, which is to say that as plant size advances (both coal and nuclear) and as capital intensity increases (nuclear), costs rise. The conclusion to be reached is that if these trends can be accepted as indicative, they are strongly incompatible with the economies-of-scale and natural-monopoly justifications for the permanent grant of official monopoly status to the established central-station utilities.

Summary

Because of high capital costs, the trailing off of cost-reducing technological innovations, faulty forecasting of demand growth on the part of utilities, and the decision to adopt commercial nuclear power, most utilities are not positioned to provide any close approximation to least-cost solutions to the nation's energy-service problems. The overly sympathetic but generally confused system of regulation, when combined with the tenacity of utility managements in attempting to save the utilities from the consequences of their faulty planning decisions, has led to policies that block or at least hinder the implementation of any creative solution to the electric supply problem that does not at the same time salvage the fortunes of the existing utilities.

In historical terms, the years since the energy crisis of the 1970s have produced a reversal of the regime prevailing up until that time. Before 1973, expansion of the size of generating plants and individual utilities was uniformly productive of falling costs. After the energy crises, high fuel costs, rapidly rising capital costs, and overoptimism in capital planning led to a situation where prices began to rise in real terms, resulting in a constriction in demand relative to historical trends, but most dramatically with respect to the dollar value of the capital that utilities were committing to an expansion of capacity. Once demand began to fall relative to capacity growth, the efficiency of utility utilization of installed capacity also began to decline, thereby driving a process of further price increases, leading to demand reductions, and to further inefficiency in plant use.

Utility commissions now must deal with utilities that have large amounts of excess capacity and poor utilization rates. The obvious solution for increasing the efficiency of capital so that electricity could be produced at a cost covered by the minimum necessary equilibrium price would be to write off large amounts of redundant and economically or technologically obsolete capital. Yet such a policy would be so traumatic that it does not appear to be a viable alternative to the regulatory commissions with their close industry ties. Under free market conditions, many of the utilities would be forced either ro write down capital voluntarily or to pass through revaluation in bankruptcy. Support from regulatory commissions, at the expense of consumers, has prevented this, while at the same time it has proved impossible for the commissions to allow utilities to finance their excess capacity completely out of higher rates, due on the one hand to the success of consumer interventions in rate cases and the price elasticity of demand—based at least in part on the increasing availability of alternative technologies for serving the energy needs of consumers.

Small-Scale Power Production and Conservation

Although the debate is typically conducted in terms of the provision of electric power, it is clear that electricity is not a final product at all but only a means to the provision of certain services (such as heating, lighting, or motive power), which may themselves be either intermediate or final goods. This distinction is important. If the electric power infrastructure can be seen as having the responsibility for providing for the services they currently supply by electricity, then the issue is not how electricity can be provided most cheaply but instead how can the demand for the services that central station electricity currently provides can be served most cheaply. Viewing the problem in this light provides an entirely new range of technological alternatives that previously have been excluded since the definition of the problem has been dominated by the vested interests. Given the availability of alternative technologies for serving the energy demand currently served by the central-station utilities, to what extent and in what manner does the current regulatory scheme prevent or hinder their free access to the market?

In theory, the central-station monopoly rests on the notion that if all demand is concentrated in the central-station provider, society at large will benefit from substantially reduced costs of service. This proposition is held with sufficient strength to create effectual barriers to the market entry of some technologies, while with perfect inconsistency ignoring the substantial share of the relevant market taken away from the central-station utilities by others. A few examples will illustrate the point.

An important element in the sales of electric utilities in the cooler regions of the nation comes from space heating. Yet though the natural-gas, coal, and fuel-oil companies split off large segments of the total load in this area, thereby presumably raising the costs of all consumers of electricity, the monopoly is not applied against these suppliers, even though large segments of the natural-gas industry are regulated by the same commissions.

A second large segment of load is lost to the utilities through self-generation by major industrial users. Under the conditions of the monopoly as it now stands, it is legal for large industrial users, or for that matter for anyone else, to generate their own electricity, even though this would entail substantially higher costs to all other consumers, in the logic of the natural-monopoly argument.

The character of these examples might lead one to think that the crux of the matter is to be found in a willingness to allow in-house generation as long as no self-generated power is sold to outsiders. Yet from the point of view of the natural-monopoly rationale, why is it not anomalous to prohibit an industrial firm with a steam electric generator from selling electricity for space heating, by private lines, over the back fence, for example, when it is acceptable to have the firm sell the steam by tubes, for the same purpose, with the same impact on the electric utilities' space-heating load?

In some small applications, it is also legal to resell electricity. An apartment-house owner with a master meter resells electricity to tenants (and who is to say that the overall rent payment does not include a markup on the sale?). Yet under existing regulation, even with PURPA, it would be strictly illegal for a farmer with a wind generator to sell a neighbor the services of a 60 watt yard light by an extension cord.

The Transmission and Distribution Monopoly

The one thread that seems to tie the examples together in some roughly logical fashion is that of access to the right of distribution, including on a larger-scale transmission. It has been argued by the utility companies, and accepted by virtually all regulatory bodies, that the official monopolies of the companies rest not alone on economies of scale in generation but on the need to avoid a functional redundancy in the creation of competing transmissions and distribution systems, which if allowed would lead, so the argument goes, to substantially higher costs to the ratepayers.[40]

In choosing the means to ensure the nonduplication of transmission and distribution (T&D) systems, however, the commissions have gone considerably further than is necessary or productive in the public interest. They have not only granted the existing central-station corporations the right to build the only T&D system within the utility's exclusive service district that would be sufficient to effect their legitimate purpose, but they have also granted the utility corporations the right of exclusive use. In this latter decision lies the error. If the T&D systems—each created through the use of the public's power of eminent domain—were to have been required to be common carriers, open to all at a reasonable toll (as are the railroads), then the situation would not exist where the utilities are free to use their more-or-less absolute control of the T&D systems to hinder or eliminate otherwise legitimate competition. Regulators, however, were no doubt comfortable in granting this absolute control over the T&D systems to the utilities because of their conviction that since a natural monopoly existed in generation, no legitimate competition could exist for the use of the T&D grid in any case. If, however, we accept for the moment that central-station generation, as it currently exists in the United States, does not constitute a natural monopoly, then to continue to allow the central-station utilities to exclude arguably more-efficient competitors by a perverse use of their legal monopoly over the T&D system constitutes the most-important barrier to reform, technological advance, and improvement in general welfare in this area. As long as the T&D systems are not available as common carriers, the utilities will have a sure mechanism for protecting their large vested interest in overvalued and technically and economically obsolete capacity at public expense.

Exclusive Service Districts

Reinforcing and cementing the transmission, distribution, and generation monopolies has been the grant of exclusive service districts to all major utilities. Except for enclaves served by municipalities, this has meant, at least before the passage of PURPA, that no other producer of electricity for sale could do business in the particular geographic area assigned to the utility by the commission. During those times when it could be legitimately assumed that the central-station utilities were the least-cost producers, it may have been acceptable to have a policy institutionalizing a monopoly of this sort. Under current conditions, all it serves to do is create economic rents for the holder of the monopoly and hinder technological advancement.

Subsidies and Immunities

In addition to the grant of official monopolies to the existing central-station utilities at the state level, the federal government has provided legal immunities and financial and R&D subsidies of great value to the utilities, in effect adopting a state technology much the same as Great Britain and Peru have state religions. The most important of these federal immunities is contained in the Price-Anderson Act, which insulates utilities with nuclear facilities from any liability in excess of $560 million arising from a nuclear accident.[41] It is generally conceded by both sides in the debate over nuclear power that without the substantial protection of the Price-Anderson Act, nuclear power would not be commercially feasible in this country because the utilities and their investors would be unwilling to bear the risk of disaster.[42]

Congress also has granted substantial tax subsidies to both coal and nuclear-based utilities that in the main have not been available, certainly not in the same profusion and magnitude, to alternative technologies. Through the provisions in the Internal Revenue Code allowing accelerated depreciation, creating the investment tax credit, and mandating "normalization" of these benefits for tax purposes, central-station utilities with large construction programs in essence have been allowed to keep the payments made by ratepayers to offset the utilities' supposed federal income-tax payments.

In terms of R&D, the relative expenditures of the federal government on central-station technology, especially nuclear power, has dwarfed its expenditures on any of the alternatives or even all of them taken together. Especially in the area of nuclear-waste disposal, the federal government has played a preeminent role in both the research area and in the creation of an institutional framework to insulate the nuclear utilities from the unknown but certainly large costs of waste transportation, storage, and maintenance.[43]

In an attempt to maintain the public's favorable impression of the economics of nuclear power relative to coal and other technologies, regulatory bodies have either ignored or minimized the costs of decommissioning nuclear facilities in their review and approval of nuclear-based capital expansion plans, assuming, no doubt, that costs of unknown magnitude should be counted as zero. This will lead—as the ratepayers in the Three Mile Island service area are currently finding—to substantial rate increases when large numbers of nuclear facilities begin to go off-line around 1995.

In addition to favorable treatment in these areas, an increasing number of state commissions, as well as the FERC, have created a means whereby utilities may charge ratepayers for facilities in the planning or construction stage long before they begin to produce any power. Under the rubric of CWIP (construction work in progress), utilities can charge ratepayers for expenses plus profit on new plants that may never be used.[44] Such treatment of plants "not used and useful" is consistent with policies that allow utilities that abandon partially constructed but demonstrably unnecessary nuclear plants to charge customers for their wasted sunk costs in those ventures and policies that require ratepayers to pay for the nonuse of idle capacity.

In considering this broad base of state and federal subsidies, two critical points emerge. The first is that given the maze of subsidies and special privileges along with the monopoly rents they have created, there is no way to untangle the true economics of the situation sufficiently to support the argument that the existing utilities are the least-cost producers of the services provided by the electricity they produce. In fact, given that the subsidies and other sorts of preferential treatment were accorded to the utilities because they were necessary to maintain viability of the utilities (as is clearly the case with nuclear power), the conclusion must be that they would have been overwhelmed by alternative, lesser-cost means of serving the same needs without them.

The second point is that only a fraction of the subsidies, and none of the privileges, are available to the small-scale, decentralized family of energy-service alternatives. If the goal is to provide the public with that changing mix of technologies and services that meet its changing energy-related needs on a least-cost basis, then there has to be a mechanism for a continual, rigorous comparison of competing alternatives. Unfortunately, the current regulatory apparatus has wedded itself to a particular technology and set of vested interests and ceased to be a protector of the public or even a neutral arbiter between competing producers.

The Response to PURPA

The passage of PURPA was to have gone a long way toward creating inter-technology competition in the energy-services markets traditionally served exclusively by the electric and natural-gas companies. In some ways, especially with respect to Title I's emphasis on the reform of rate structures, PURPA has been the catalyst for a good deal of activity, much of it progressive, on the part of state commissions (though the progressive state commissions seem to be located in the states with utilities not suffering from large amounts of effective excess capacity).[45] In the area of small-scale power production (the Title II reforms), movement has been less evident, and substantial resistance has been generated at certain critical points where regulators wanting to minimize discord with the utilities, and utility managements arrayed to protect the fruits of their embattled monopoly, have been able to comply with the letter of the law while thwarting its clear intent.

The first of these sticking points is in the area of rates. The rules issued by FERC in support of PURPA require that utilities purchase the power produced by qualifying small-scale facilities (QFs). The dollar value of the rates to be paid are to be based on the energy and capacity costs avoided by the central-station utility in buying power from the QF. The FERC rules were explicit in requiring that these avoided-cost rates were to be divided into an energy-cost (utility fuel savings) component and a capacity component (capital costs saved).

Unfortunately, in addition to arguing that the QF-produced power was in general not serviceable in terms of the technical characteristics of electricity produced, many utilities objected to paying any capacity credit for any QF power on the grounds that since they had an excess of generating capacity of their own, they saved no capital costs through buying power from small-scale producers. This effective elimination of the capacity credit made it uneconomical for the majority of potential QFs to produce power for sale to the utility, and since sales to any but the central-station utilities were prohibited under the states' exclusive-service district rules, the bulk of potential small-scale production was prevented from coming on line.

This is a perverse result in terms of the logic of the market and community welfare for several reasons. First, since utility rates charged customers reflect the large amounts of excess capacity at very high capital costs, consumers might well be better off in the long run to develop competitive alternatives to the overcapitalized central-station utilities since unit costs for the more-advanced alternatives are falling over time while utility rates are slated to continue their rapid rise.[46] Second, allowing the utilities to reject capacity credits on these grounds means that regulators are allowing them to use the fruits of their planning errors as a defense against the legitimate response to those errors in the form of the creation of alternative sources of supply. The effect on the QFs is compounded by the fact that the QFs can have no legal access to the market other than through the existing utilities; to block their access along that route is to eliminate them. If the utilities are to be allowed to refuse to deal with small-scale producers under these conditions and in the face of the congressional determination that the development of small-scale power production is in the national interest, then the only other alternative is to allow the QFs to seek markets elsewhere. The public has no interest in protecting the utilities' investment in economically superfluous capacity; to allow the utilities to refuse to deal while at the same time insisting on the preservation of the right of the utilities to exclusive access to the market is clearly not in the public interest.

The utilities have also objected to capacity credits on the grounds that individual solar, wind, or hydroelectric facilities are not capable of supplying a sufficiently constant power flow to be classed as firm capacity; that is, they could not be substituted for reliable thermal generators and therefore do not meet the basic qualifications for capacity credits. However, efforts to obtain regulatory recognition for groups of diverse qualifying facilities organized so as to produce firm power in combination have been unsuccessful.[47]

These measures have not defeated the utilities and unsympathetic regulators. By means of high interconnection charges and high backup rates for QFs, utilities have been successful in further discouraging small-scale production. The issue of the fees for backup power has been especially perplexing given the element of contradiction involved when utilities justify the high rates on the grounds that they would have to purchase new capacity to provide the seldom-used backup power, yet when the issue of capacity credits arises, their arguments seem to take just the opposite tack.

Complementing utility resistance in these areas have been two challenges to the legality of PURPA in the federal courts. The first case was brought by the Mississippi Public Utilities Commission and Mississippi Power and Light, challenging the constitutionality of Titles I and III and section 210 of PURPA on the grounds that it violated states' rights as protected by the Tenth Amendment. Mississippi was successful at the federal district court level but lost on appeal to the U.S. Supreme Court.[48] In the interim sixteen states slowed or suspended compliance with the PURPA mandates, putting them years behind the compliance schedule.[49]

The second challenge, aimed directly against the FERC rules promulgating the avoided-cost standard, was successfully maintained by American Electric Power in the District of Columbia federal circuit court.[50] The impact of that decision (pending FERC's request for a rehearing and possible appeal to the Supreme Court, a procedure that might take up to two years) is to cast doubt on the continued viability of all QF operations not currently under firm contracts. This sharply increased risk, together with already high interest rates and the collapse of federal funding under Title IV of PURPA, have been effective in sharply limiting the ability of small-scale power producers to compete.

In sum, the impact of PURPA has been to stimulate interest in industry reform and to breach the first line of regulatory and utility barriers to the creation of an effective set of alternatives. With the diminished commitment to utility and energy reform on the part of the current national administration, however, further advance has been successfully slowed or stopped altogether except in states where, because of local economic and political conditions, commissions are committed to make progress in these areas on their own (for example, in California, Vermont, and New Hampshire).

An Agenda for Reform

Many of the problems afflicting the electric utility industry stem from a single root: the disastrous capacity-planning experience during the 1970s in which the utilities committed themselves to truly awesome capital expenditures on large amounts of highly capital-intensive generating capacity at a time when the country was heading into a period of very high interest rates, poor economic conditions, large-scale shifts of industry from region to region and abroad, and a high level of economic uncertainty. Under these conditions, generation planning should have turned toward a greater emphasis on short-term flexibility in capital planning: smaller plants, shorter construction lead times, emphasis on capital-savings means of meeting generation needs, and a program of integrating the utilities' capital-expansion plan into the community's need for economic viability. The result of the consequent dramatic overexpansion of large-scale central station capacity—expansion far beyond the ability of the public to assimilate it profitably, given the cost of the capital to the utilities—has been to create a large-scale financial crisis for the utilities and many of their customers.

Under these circumstances, utility managements have been led to resist all measures of community energy planning that interfere in any way with the recoupment of costs plus return from this superfluous capacity. Efforts to reform rates to reduce demand peaks were, and are, resisted because they would call attention to surplus capacity; efforts to promote conservation, insulation, and increased appliance efficiency have the same effect and stimulate the same response. The creation of small-scale power alternatives, and especially cogeneration, promises to reduce utility sales and to snowball into a substantial challenge as utility rates rise as yet further superfluous capacity is brought on-line, and as reduced demand relative to installed capacity leads to higher unit rates because of the need to spread a greater volume of fixed costs over a relatively smaller number of kilowatt-hours sold.

The goals and problems of the utilities are but one side of the energy coin, with the needs of the public, including industry and state and local governments as well as the residential sector, occupying the other. The basic requirement of each of these sectors is to increase energy efficiency and to decrease cost. This means that at sharply higher prices for a kilowatt-hour of electricity, it is necessary to find ways to use fewer kilowatt-hours (conservation), to identify production alternatives that will produce kilowatt-hours at a lower cost, and to develop other technologies for providing substitute means for accomplishing functions heretofore served by kilowatt-hours of electricity.

The economic pressures on the ratepaying public to find solutions along these lines are substantial, yet as far as electricity goes, an important part of the impetus toward the creation of less-costly alternatives is not due to the basic increase in the necessary and efficient costs of supplying power in the traditional fashion but to glaring inefficiencies and waste in capital planning by utilities.

The objectives of most utilities—especially those with large construction projects (especially nuclear) and substantial amounts of excess capacity—conflict with legitimate public needs. The four years since the passage of PURPA in 1978 have been a period of attempted adjustment among ratepayers, utility commissions, the utilities, and alternative producers. Over this period there has been some progress, to be sure; nevertheless, the basic problem still remains and is getting worse in many places due to continuing expansion on the same model. In the end, the public will allow the utilities to drain off hundreds of billions of dollars in rate increase to pay for unnecessary capital. What is needed is effective reform of the institutions and processes controlling the provision of energy services in the functional areas now generally served by the central-station utilities. This agenda can be met in four ways: pricing, regulation, industrial organization, and alternative technologies.

Pricing

Three main purposes could be served through appropriate rate design: the generation of utility revenue in a fashion serviceable to public goals, the conveyance of information to users about the efficient use of resources, and the creation of an equitable system for distributing costs among users. In mandating that rates be based on the cost of serving particular classes of customers and in requiring consideration of time-of-day (TOD) and seasonal rates, PURPA addressed in part the equity and efficiency considerations. And although cost-of-service, TOD, and seasonal pricing are moves in the right direction (for those utilities adopting them), they tend to focus on short-term efficiency relative to the existing generation set. What is necessary is some mechanism to provide the public with signals to guide them in their decisions that affect capital planning through the growth rate in demand.

As the pricing system is currently organized, the public pays rates sufficient to cover the costs of the existing system. But since new capacity is much more expensive than the average of existing capacity in the rate base, the coming on-line of new capacity is likely to result in a substantial jump in rates (20 to 40 percent increases in rates are not uncommon when large units owned by relatively small utilities come on-line).

In terms of the character of the information contained in the rates that ratepayers were facing before the completion of the new capacity, consumers were led to believe that they could increase their consumption (thereby using up existing excess capacity and bringing the date nearer when new capacity would have to be brought on-line) at the old price. Under this pricing scheme, there is no price signal whatsoever to the consumers that their extra consumption will cause a large increase in their costs per unit at some point in the future, dependent on how quickly what is left is used up. Because they have no warning, they have no opportunity to make economically efficient choices until after the decision their choices could have had an impact upon is already a completed fact.

An alternative is a pricing scheme based on dynamic long-run, marginal-cost pricing (DLRMCP), which would supply ratepayers with a pricing signal encouraging them to use idle capacity when it is plentiful but to conserve on use when an expansion in demand would create the need for an early addition to expensive generating capacity. The mechanism for accomplishing this purpose is to include in rates an element reflecting the opportunity cost (or shadow price) of existing excess capacity at current rates of growth in demand. The existence of large amounts of excess capacity and low rates of demand growth would result in a DLRMCP price element that was quite low, indicating that consumers could continue to expand demand and not create the need to build new capacity anytime soon. On the other hand, small amounts of excess capacity at high rates of growth, with a large prospective increase in generating costs in the offing, would entail a charge set at a high and increasing level that would rise in time so that the total rate at some point would approximate the rate that would occur when the new capacity is built and brought on-line. In this way the utility would know that when demand is large enough at rates that would approximate the rates necessary to support new capacity, it is time to build that new capacity. Consumers would be led to conserve on all uses that they would not consider necessary at the level of rates required to support a new plant. In this way both conservation and capital planning are jointly optimized, and both consumers and utility managements receive a constant flow of appropriate price and demand signals over time, which must lead to better decision making and the elimination of price shocks to the consumers and capacity-demand problems for the utilities.

In one of the most-peculiar arguments of recent times, many utilities objected to the implementation of DLRMCP on the grounds that it would have a tendency to deliver them excess revenues over actual costs, thereby supposedly violating the rule of regulation that utility revenues must be just sufficient to cover costs plus a reasonable rate of return. Although it is quite clear that such a situation might occur at times when the need to build new capacity is imminent and DLRMCP charges are high, the difficulty is illusory. The excess revenues can either be used to provide rate-based loans to enable ratepayers to implement cost-effective conservation measures to help defer the new plant, or the utility could bank the excess as contribution in aid of construction (CIAC) for use in helping to offset the cost of the new plant when it came to be required. In either case, the proposed expenditure is a cost to the utility, and no excess in rate of return would occur.

From the point of view of the utility, the real objection to DLRMCP is that it has the vice of its virtues. Decreases in consumer costs over the long run are seen as losses in revenues, and the predicated pacification of the present frenetic pace of new-plant construction is seen as a threat to utilities that have to maintain their construction programs at high levels in order to avoid tax problems associated with accelerated depreciation.

Regulatory Reforms

Five measures may be useful: the strict application of the used-and-useful standard in dealing with excess capacity, recognition of the end of natural monopoly in generation, common-carrier status for the T&D systems, consideration of a federal power grid, and the creation of full-service energy utilities.

The Used-and-Useful Standard: It is a long-held and fully established rule of ratemaking in regulated industries that regulators can allow in the rate base only property currently used and useful in the provision of jurisdictional services.[51]

> As of right safeguarded by the due process clause of the Fifth Amendment, appellant is entitled to rates, not per se excessive and extortionate, sufficient to yield a reasonable rate of return upon the value of property used, at the time it is being used, to render the services. [Citations omitted.] But it is not entitled to have included any property not used and useful for that purpose.[52]

The major roadblock to attaining the necessary reforms in the electric power industry is the presence of large amounts of such property "not used and useful." A firm hand in the exclusion of such capacity from the rate base would rapidly reform utility capital planning. All new planning would be under the clear realization that utility management, always jealous of its investment planning prerogatives, would be held responsible for errors of planning, forecasting, or judgment, for who else—certainly not the public with no knowledgeable connection to the utility planning process at all—should be responsible except the people who insist on the prerogatives of decision making. If, on the other hand, the decision is to force consumers instead of utility stockholders to bear the risks, then the preferred organizational form of the utilities is state or cooperative ownership. If this is unacceptable, then given private ownership, the investor bears the risk, even with regulation.

A second difficulty with management's making the decisions but not being held responsible for adverse consequences is that moral hazard enters the equation. If decision makers bear none of the costs of risk, then they will tend to adopt riskier plans than would otherwise be the case. This is most evident in the admission that without the Price-Anderson Act's shifting of risk away from utility stockholders, they would have picked some other technology in preference to nuclear power. Or in terms of safety investments, if the utility's liability for an accident is limited at $560 million and the cost of prevention is, for example, $600 million, then even if the damages to the public would be $10 billion, the company would have no financial interest in making the investments. If there is no penalty for having large amounts of excess capacity, why invest time and money in trying to improve methods to avoid it? If the costs to the utility of sustained outage are large but the social costs of excess capacity that can be brought home to the utility are negligible, then rational planning will always produce excess capacity. In order to produce publicly serviceable planning from utility managements operating on private enterprise principles, regulators have to create an incentive structure that will lead managers to enter the true costs of excess capacity to society into their own corporate profit-making calculus. The only way that this can be done without having the public take over the planning function is to force the utility to bear the full cost of its planning errors.

Natural Monopoly: The question of the natural-monopoly status of utility-generating systems has been investigated in detail. Given the conclusion that the existing utilities are not the least-cost producers of the relevant energy services, the natural-monopoly basis for the grant of official monopoly ought to be dropped (along with the monopoly and exclusive-district grants themselves). A more-pragmatic rationale—for instance, the control of market power—ought to be substituted instead.[53]

Without exclusive service districts but with a continuation of rate regulation for all utilities, competition would allocate customers to those utilities that could supply service most cheaply. This result is clearly in the public interest.

Common Carrier Status for T&D Grids: The argument that the elimination of exclusive service districts would lead to the creation of duplicative T&D systems could be avoided by the requirement that T&D grids built with the power of eminent domain be common carriers. This would facilitate not only competition between large utilities but also allow small producers access to local markets, thereby eliminating the existing bottleneck in that area.

A Federal Power Grid: One partial alternative to state-mandated common-carrier status is the creation of a federal power grid to act as a common carrier. This would eliminate the existing transmission monopoly held by the large utilities, which is used all too often as an anticompetitive device to suppress nascent municipal utilities.

Full-Service Utilities: With the growing importance and cost-effectiveness of conservation and other alternative means of providing energy services, the need for an institutional focal point for information, expertise, and financing to facilitate efforts in this area becomes evident. In several jurisdictions (California, for example), responsibility for making loans available to ratepayers for conservation or solar improvements has been assigned to utilities. In many areas municipal utilities and federal agencies like the Tennessee Valley Authority have been active in providing energy audits, making loans for improvements, certifying and supervising solar energy and insulation contractors, and involving themselves in alternative generating technologies. In some places, cities have gone further, coordinating local land-use planning, building regulations, and other police-power controls, to produce an energy-efficient and cost-effective environment.

Since all of the services provided by the full-service utility are cost compensated and enter the rate base (if they involve capital expenditures), there can be no legitimate financial argument on the part of existing utilities against offering this fuller range of services. Utilities, however, typically argue that they are neither knowledgeable about nor equipped to manage these service offerings. It must be concluded, however, that the objection is a hollow one when the utilities' demand to be allowed to diversify into nonutility-related enterprises is considered at the same time.

For situations where the services of a full-service utility have been determined to be in the public interest and the existing local utility is adamant in refusing to supply them, it may be time to consider replacing the utility with a more-tractable competitor or with a municipal or cooperative enterprise.

Industrial Organization

Forty-seven years after the passage of the Public Utility Holding Company Act in 1935, a strong movement among utility interests has set in motion an effort to outright repeal. In support are the White House, a unanimous Securities and Exchange Commission, the Department of Energy, and allied interests in both houses of Congress (notably absent is the Department of Justice).[54] The purpose of repeal is cloaked in vague generalities about regulatory burden and outdated legislation, but the main point seems to be to allow utility diversification into more-profitable areas.

Three problems are associated with allowing the industry to diversify freely. First, the mixture of regulated with unregulated activities dramatically increases (as it did before 1935) the burden of determining legitimate rates. Second, the antitrust implications in terms of suppression of intertechnology competition are serious.[55] And third, it is not at all clear that the utilities would be helped by the greater profitability of their acquisitions, rather than being decapitalized by them in what has become a standard motif in a number of other industries. A fourth problem, discussed in some detail by later-to-be Supreme Court Justice Louis D. Brandeis under the heading "The Inefficiency of the Oligarchs," was that there are inefficiencies inherent in diversification outside the range of the expertise of top management, or as he put it as his second law of human limitations, "A man cannot at the same time do many things well."[56]

To allow the electric utilities to diversify either into competing technologies, or into unregulated fuel or equipment-supply industries, or into the ranks of their customers, or in the end to escape the scene of their planning debacle with as many of the remaining assets as possible leaving but a hollow shell behind, cannot be in the public interest. In fact it would seem more to the point, if the point is indeed to encourage a resurgence of productive intertechnology competition, to force the utilities to divest the holdings they already have in coal mines, technologically unrelated R&D companies, natural-gas subsidiaries, solar contractors, and other similar enterprises.[57]

Divestiture along these lines, together with other reforms, would be effective in breaking the stranglehold the existing utilities now have on the market and the potential for creative enterprise.

Alternative Technologies

In terms of encouraging the development of cost-effective alternatives, the most-necessary reform is the elimination of the anticompetitive effects of existing regulation and the restrictive activities of central-station utilities. Allowing sales from small producers directly to consumers without price regulation, by common-carrier T&D grids while regulating the rates of larger alternative producers' under a regime of free consumer choice of supplier, would be sufficient to create incentives to the development of wind, solar, biomass, small hydro, and cogeneration.[58] Favorable tax incentives, technical assistance, and other incentives would be useful in speeding the acceptance of these technologies by the market but are probably not essential. In its report on energy alternatives, the Solar Energy Research Institute (SERI) estimated that with the full implementation of cost-effective energy conservation (*not* including potential savings from the use of solar water or space heating, photovoltaics, or on-site cogeneration), demand growth facing the central-station utilities could be reduced to 0.4 percent per annum. Even modest reductions below *Electrical World's* forecast 3.3 percent growth rate would allow substantial savings to be made over the $348 billion investment in new capacity the Edison Electric Institute estimates as being required over the next eight years.

> Electric utility equipment now operating or under advanced construction could sustain a growth in demand of 0.1–0.3 percent per year for 20 years, even if all the obsolescent plants and most oil and gas facilities are retired.... A demand growth of 1–2 percent could be sustained given the economic use of cogeneration in the six major steam consuming industries.[59]

Such is the potential for a rigorous but cost-effective application of alternative technologies, combined with the judicious use of existing excess capacity over the next twenty years. The means to accomplishment are technological innovation, an end to unnecessary monopoly, and regulatory reform. For the existing utilities, the motto of Sawhill, Texas—lead, follow, or get out of the way—may, for their sakes, as well as the public's, be worthy of serious consideration.

Notes

1. With the Wisconsin Public Utility Law (designed by J.R. Commons). See Keith M. Howe and Eugene Ramussen, *Public Utility Economics and*

Finance (Englewood Cliffs, N.J.: Prentice-Hall, 1981) for a brief discussion of the history of early regulation.

2. "Overlay" because the structure of municipal regulation under franchise was allowed to continue in many states without direct control of rates.

3. Sheldon Novick has argued that state-level regulation was seen by the more-insightful utility leaders (notably Insull) as necessary and useful for the industry. On the one hand, state regulation would tend to block the movement toward state ownership, while on the other it would protect against competition and, under Supreme Court direction, guarantee a fair rate of return on investments in plant and equipment. A fair rate of return on investment seems like a minimal sort of guideline. In practice, the effect was to permit power companies to charge rates sufficient to pay for the physical plant they built regardless of the motive or the prudence of the construction and to encourage new and still-larger investments. Large power plants could be built even though they would be only partially used; existing customers would pay for the expansion, while the huge new surplus capacity would be used to solicit new customers for large blocks of power at very low prices. State regulation would ensure that the companies could charge enough to any class of customers to bring themselves a fair return on investments.

4. "Samuel Insull's holding company, Middle West Utilities Company, provided utility services through its operating subsidiaries to more than 5300 communities in 32 states, mostly in nonmetropolitan areas. In 1932 Samuel Insull was president of 11 power companies, chairman of 65, and director of 85." See Davis Morris, *Self-Reliant Cities* (San Francisco: Sierra Club Books, 1982), p. 41.

5. James C. Bonbright, "Should the Utility Holding Company Be Regulated? *Public Utility Fortnightly,* Feb. 19, 1931, p. 200.

6. Ibid., p. 201.

7. For a good general discussion of this, see A.A. Berle, Jr., "Subsidiary Corporations and Credit Manipulation," *Harvard Law Review* 41 (May 1928): 874, 875–876.

8. Edward T. Myers, "The Great Railroad Robbery: Disinvestment," *Modern Railroads* (September 1974): 69; and Barry Bluestone and Bennett Harrison, "Why Corporations Close Profitable Plants" *Working Papers* (May– June 1980): 15–23.

9. Berle, "Subsidiary Corporations," p. 875.

10. This was done through the Federal Power Act of 1935. For further discussion see Howe and Rasmussen, *Public Utility Economics,* pp. 44–45.

11. Technical Information Center, *Nuclear Reactors, Built, Being Built, or Planned* (Washington, D.C.: U.S. Department of Energy, 1982), pp. 7–11.

12. Charles Komanoff, *Power Plant Cost Escalation: Nuclear and Coal Capital Costs, Regulations, and Economics* (New York: Konanoff Energy Associates, 1981), p. 262.

13. The reason that uranium prices have not maintained the high levels of the mid-1970s is not because of the discovery of vast new domestic supplies but because of the collapse of prospective demand, especially after the Three Mile Island accident.

14. Note, "The Constitutionality and Effectiveness of the Electric Utility Provisions of the Public Utility Regulatory Policies Act of 1978," *George Washington Law Review* 47 (1979): 787, 800.

15. Ibid., p. 795.

16. Under declining block rates, as consumption increases, the price per unit consumed falls.

17. See the discussion in James C. Bonbright, *Principles of Public Utility Rates* (New York: Columbia University Press, 1961), pp. 20–22.

18. Cogeneration means the use of the process steam to produce electricity, or, the other way around, the use of steam from the generation of electric power to provide process heat. For an extended discussion of the possibilities, see EPRI, *Cogeneration and Central Station Generation* (Palo Alto: EPRI, 1981). Also see the discussion of "cascading" heat use; in Morris, *Self-Reliant Cities,* p. 189.

19. See Morris, *Self-Reliant Cities,* pp. 143–152 and passim for a brief description of utility efforts to block small-scale power production.

20 Sections 133, 208, and 210 of PURPA (PL 95-617, 16 USC 2601 et seq.), and regulations 45 FR 12214 (2/25/80).

21. Iowa Conservation Commission, *Iowa's Low-Head Dams: Their Past, Present, and Future Roles* (Ames: Iowa Conservation Commission, 1979).

22. Allowing the dispatch means the use of that combination of generating plants that most cheaply meets total grid load, with transactions between utilities being controlled through an established grid-wide set of accounting rules and pricing principles. The system under which plants are dispatched centrally is known as economy dispatch. For a detailed discussion of these issues, refer to W.S. Nelson, *Midcontinent Area Power Planners* (East Lansing: Michigan State University Public Utilities Studies, 1968).

23. See, for example, the 1981 annual report of American Electric Power, especially pp. 34–35 on subsidiaries.

24. C.E. Ferguson, *Microeconomic Theory* (Homewood, Ill.: Irwin, 1969), p. 255.

25. Richard T. Ely, *Monopolies and Trusts* (New York: Macmillan, 1912), p. 62.

26. Bonbright, "Should the Utility Holding Company Be Regulated?"

27. Ibid., p. 13.

28. Richard Schmalensee, *The Control of Natural Monopolies* (Lexington, Mass.: Lexington Books, D.C. Heath and Company, 1979), p. 3. Kahn is in agreement with Schmalensee on the cost question though he retains the notion of natural monopoly based on the single, more-efficient plant. See

Alfred E. Kahn, *The Economics of Regulation* (New York: Wiley-Hamilton, 1971), 2:120, 121, 124–125.

29. Schmalensee, *Control,* pp. 3, 5.

30. The most-recent figures show excess capacity over peak demand nation-wide to be 30.6 percent for 1980, with the 1981 estimate at 33.3 percent and the 1982 projection at 33.4 percent, rising to 34.4 percent in 1983. *Electrical World,* 32d Annual Electrical Industry Forecast (September 1981), p. 84.

31. In terms of the logic of both the competitive market and long-established precedent in rate regulation (see *Symth v. Ames* 169 U.S. 466) to the point that "excess or fictitious capitalization" may legally be recovered from the public through regulated rates), capital not "used and useful" should not be allowed in the rate base to be passed on to the consumers in terms of higher costs.

32. AEP, for example, with very little, if any, fuel-oil-fired based-load capacity currently has 48 percent excess capacity over peak and is going ahead with an extensive program of further construction of new capacity. *Power Line* (July 1982): 6. In Iowa, with no base-load fuel-oil facilities, several of the seven major utilities have excess capacity in the range 35 to 53 percent, with new construction of base-load capacity in the works. See, for example, the prepared Testimony of Richard P. Cool in Iowa State Commerce Commission Docket No. RPU-81-5.

33. The evidence is clear for coal plants. Marie R. Cosio, "Why Is the Performance of Electric Generating Units Declining?" *Public Utility Fortnightly,* April 29, 1982, pp. 25, 27. For nuclear plants, it is more difficult to find data. The best-available data seem to indicate that the upward trend in the size of nuclear plants between 1970 and 1979 has resulted in a mild average increase in heat rates over the same period.

34. Komanoff, *Power Plant Cost Escalation,* p. 317.

35. Load factor is defined as the ratio of kilowatt-hours sold in a particular period to the peak load times the number of hours in the period. For example, if the peak demand is very high when compared to the typical off-peak load, the load factor will be very low. Low load factors indicate inefficient use of capital for a utility because the capital stock necessary to meet peak demand under conditions of low load factor would be idle the better part of the remainder of the time. If the peak were to be smoothed out, a much smaller amount of capacity would be sufficient to serve the same volume of demand in kilowatt-hours, giving a much higher rate of utilization.

36. *Electrical World* (September 1981): 84. Load factors computed from columns 1 and 3 of the table.

37. In Iowa, for example, the Iowa Commerce Commission, in making rate-of-return decisions, penalizes utilities for excess capacity based on the amount of excess capacity over peak. This practice encourages the utility to resist any

pricing reform that would help rationalize the use of capital. Decisions and Orders in Docket No. RPU-80-65 and RPU-81-8 (March 3, 1982).

38. Komanoff, *Power Plant Cost Escalation,* p. 250.

39. Ibid., p. 255.

40. For a strong presentation of the case that duplication of distribution facilities when it also entails strong competition to entrenched utility monopolies is cost effective from the point of view of ratepayers, see the excellent work by Richard Hellman, *Government Competition in the Electric Utility Industry* (New York: Praeger, 1972), esp. chap. 2 and the case study of Maquoketa, Iowa, p. 269-273.

41. This would mean, for example, that if a major nuclear accident were to occur that resulted in a rather modest 15.6 billion in damages to the public, the utility would have a legal obligation to compensate the losers for only 10 percent of their losses, with the remainder coming out of the pockets of the irradiated themselves. A challenge to the act on constitutional grounds was taken to the U.S. Supreme Court in 1981 where the act was upheld. (*Duke Power Co.* v. *Carolina Environmental Study Group, Inc.* 438 U.S. 59 (1978)).

42. Since the determination of least-cost production from the point of view of the regulatory commissions should certainly include both the direct and the indirect costs to the public, it is difficult to imagine how a legitimate determination could have been made and maintained over the last ten to fifteen years of experience with nuclear-power plants that they constitute the least-cost means of serving energy needs. Additionally, Canada's Royal Commission on Electric Power Planning estimated that the risk of a meltdown-level accident in one or more of Toronto's twelve commercial nuclear plants over the period of their thirty-year life spans is 3.6 percent, give or take a factor of five. Such an accident, were it to occur in conjunction with winds blowing in the direction of the city, would result in large casualties and require permanent evacuation of the city. Lawrence Solomon, *Energy Shock* (New York: Doubleday, 1980), pp. 179, 299.

43. The bill recently passed by the U.S. Senate, the National Nuclear Waste Policy Act (S. 1662), even goes so far as to exempt the site adoption process from the protective requirements of the National Environmental Policy Act.

44. For a detailed but readable discussion of CWIP and related issues, see the National Consumer Law Center, *Materials on Construction Work in Progress* (Washington, D.C.: NCLC, 1981).

45. For example, in New England with its economically obsolete fuel-oil capacity and California with new base-load natural-gas facilities, which it is illegal to operate.

46. For example, LILCO's Shoreham nuclear facility is coming in at $3,000 per installed kilowatt, or approximately twelve times its planned cost. It is estimated that when the plant comes on-line, it will result in a sudden 30 percent increase in rates. "Rate Shock," *Wall Street Journal,* August 11, 1982.

47. See the prepared testimony of David Osterberg and Michael Sheehan, *IN RE: Iowa State Commerce Commission Rules Regarding Rates for Cogeneration and Small Scale Power Production,* Docket No. RMU-80-15; January 25, 1981, for detailed arguments in favor of one such proposal.

48. *Federal Energy Regulatory Commission, et al., Appellants,* v. *Mississippi et al.,* No. 80-1749, decided June 1, 1982.

49. National Association of Regulatory Utility Commissioners, *Bulletin,* July 26, 1982, p. 9.

50. "Appeals Court Strikes Down Central Portions of PURPA Regulations," *Solar Law Reporter* (January–February 1982): 738. The court also held that before it can order a recalcitrant utility to interconnect with a qualifying facility, FERC must meet the hearing and notice requirements of section 210 of the Federal Power Act, which for most small QFs would be fatally burdensome. Ibid., p. 739.

51. The jurisdictional service means, for example, that a utility with a power plant that was used only to supply electric power to a neighboring utility could not be included in the rate base charged to the owning utility's own ratepayers.

52. From the U.S. Supreme Court decision in *Denver Stock Yard Co.* v. *U.S.,* 304 U.S. 470, 475 (1937).

53. If the cost-of-service pricing principle were to be given a geographic as well as a customer class dimension, then the end of exclusive service districts would not lead to socially disoptimal raiding of one utility's territory by another, except in cases where a second utility could provide as good or better service at lower cost. In this case, such an action would not be socially disoptimal.

54. Corie Brown and Barbara Bink, "Special Report: The Movement for Repeal of the Public Utility Holding Company Act," *Public Utility Fortnightly,* May 13, 1982, pp. 42–45.

55. There is already reason for concern about the penetration of the oil companies into the evolving photovoltaics industry (not to mention coal, uranium, natural gas, and synfuels), where the top three firms, controlling 77.4 percent of the market, are subsidiaries of major oil companies. Ronald Wilcox, "Oil Companies and Photovoltaics: A Potential Monopoly?" *Solar Law Reporter* 3 (November–December 1981): 706. The utility industry could be an important element in too many possible combinations of interests whose amalgamation would run against the general welfare and public purpose of this area for the act to be withdrawn on the basis of such flimsy arguments as are currently being advanced by its supporters. Who would feel comfortable, for example, were the utilities themselves to be acquired by the great oil companies?

56. Louis D. Brandeis, *Other People's Money,* (New York: Harper and Row, 1967), p. 137.

57. One of the discoveries made this year by intervenors and commission staff was that Iowa Public Service Company (one of the seven large investor-owned utilities serving Iowa) had been paying its wholly owned and unregulated coal-mining subsidiary substantially above market value for its coal. In a more-complicated holding-company structure, this might well have been impossible to detect, as commissions found before 1935. For the details of the IPS case, see Iowa State Commerce Commission, Iowa Public Service Company Request for Revision of Rates, Docket No. RPU-80-65 and RPU-81-8 (1981).

58. Small producers here is in the sense of neighbor-to-neighbor size.

59. Solar Energy Research Institute, *A New Prosperity: Building a Sustainable Energy Future* (Andover, Mass. Brick House Publishing Co., 1981), pp. 327-288 and chap. 4 generally.

7 Implementing Community Energy- Management Strategies

Beverly A. Cigler

The traditional list of local government needs and priorities was altered in the mid-1970s when local officials became painfully aware of the important fiscal, social, environmental, and political impacts of inefficient energy use, rising energy prices, and fluctuating energy supplies. The Arab oil embargo of 1973–1974 marked the beginning of the realization that energy was a significant cause of the rising cost of delivering public services and a growing limit on the choices available to local government in meeting public needs and demands. Later events—natural-gas interruptions in the winter of 1977–1978, gasoline shortages in 1979, the doubling of gasoline prices in 1979–1980, an Iranian oil shutdown in 1979, the near disaster at a nuclear plant at Three Mile Island near Harrisburg, Pennsylvania, and, most recently, double-digit inflation and a war in the Persian Gulf—contributed to placing energy center stage as a major national issue.

Local governments have been particularly hard hit by the nation's energy problems. Energy management must be included in the planning and program decisions of local governments if their fiscal, social, and economic goals are to be met. A minor budget item ten years ago, energy has grown to a major local government expense. An International City Management Association (ICMA) study found energy costs to be the second largest item in the budgets of nearly 60 percent of our cities and more than 40 percent of our counties. Only personnel costs exceed energy costs.[1] Several studies of energy consumption, cost, and expenditure patterns of communities have revealed that about 85 cents of every dollar spent on energy leaves the local economy. This compares to between $0.75 and $1.25 returned for every dollar paid in federal taxes. Many communities spend $1,000 per capita on energy.[2]

Three Stages of Local Energy Involvement

Local involvement in energy matters has occurred in three stages. The years between 1973 and 1979 marked a formative period of education, pilot projects, and examination of local potential. The macro approach of federal mandates, decontrol, and deregulation dominated, with attention focused on auto-efficiency standards, establishment of mandatory speed limits, and building temperatures, among other federal efforts. Scattered community leaders calling for local solutions to the energy problem were largely ignored.[3] Several demonstration projects funded by the Departments of Energy (DOE) and Housing and Urban Development (HUD) supported the local energy role, but these were fragmented and small in scope.[4] There were few adoptions spawned by these efforts and little documentation of results.[5]

This early period, however, did demonstrate the enormous potential of local energy programs. It was realized that local governments could make significant contributions toward increasing the energy efficiency of the nation by promoting, regulating, or providing incentives for energy efficiency. The reasons justifying a local role in energy policy and management include: (1) energy management as a fiscal tool and cutback management strategy; (2) conservation and renewable-resources development as stabilizers of the economic base of a city or county; (3) a local role in ameliorating attitudinal and institutional barriers to energy efficiency and production options; and (4) a means of acknowledging the unique differences in local climate, geography, economic and political institutions, and physical development among communities.[6]

While an understanding of the local role in energy planning emerged in the early period, only a few cities, such as the widely-publicized cases of Davis, California, and Seattle and Portland, Oregon, actually developed and implemented comprehensive energy programs.[7] For the most part, the country was complacent when oil supplies were reestablished after the Arab oil embargo and when the price of oil actually fell relative to the consumer price index between 1974 and 1978.

The second era of local energy involvement began in 1979 when complacency was ended and Americans became accustomed to hearing startling energy news relating to the toppling of the shah of Iran, OPEC's ability to double world oil prices, the seizure of American hostages in Iran, and a serious nuclear accident in Pennsylvania. A new terminology emerged within hundreds of local governments, *community energy management* (CEM), as leaders and citizens mobilized to develop comprehensive energy programs.[8]

Local officials' consciousness of energy-management potential led to the development of more than a hundred comprehensive energy-management plans and other local energy options.[9] Energy efficiency became widely recognized as the nation's largest and cheapest source of additional energy. Reducing energy waste remains the best near-term option for achieving a reduction in oil imports at reasonable cost, speed, and certainty in the 1980s. Research estimates suggest that between 40 and 50 percent of the energy used in the United States could be conserved through economically justified measures that could be implemented on a wide scale within the next fifteen years.[10]

Hundreds of guidebooks for community energy management were produced and distributed by governments at all levels, public-interest groups, professional organizations representing local governments, and academic researchers during the second period.[11] Local governments appeared ready to embark on serious efforts toward using energy programs as cost-avoidance strategies, helping overall national energy prospects.

The second era in local energy management experienced a major jolt with the election of President Reagan in 1980. Fiscal constraints at all levels of government have forced many local governments to discontinue, postpone, and cut back local energy-management programs. The Reagan administration's energy policies rely almost exclusively on market forces, eliminating most of the technical and financial-assistance programs available to local governments to develop and implement community energy programs.[12] The Reagan administration, for example, proposed a 1983 budget that would eliminate most conservation programs entirely and reduce the total amount spent on conservation by 97 percent. Most information development and dissemination activities have been eliminated to date. While Congress has generally appropriated far more for conservation and renewable-resources programs than has been required by the Reagan administration, funding cuts have still ranged between 50 percent and 100 percent, depending on the program, between 1980 and 1983.[13]

A third era in local energy involvement has emerged, marked by a sudden shift in national priorities. Local governments will have to work alone if they are interested in comprehensive community energy-management programs. Interestingly, the shift of authority to subnational governments and the private sector, the basis of the Reagan domestic programs, fits well with the CEM philosophy. Federal support for local efforts is no longer likely, but there has been some help for local governments during the transition. DOE recently has funded projects that help synthesize knowledge and experience with CEM so that other communities can marshal knowledge and avoid repeating unsuccessful approaches to energy management.[14]

In early 1981 the National Community Energy Management Center was created by a joint agreement among HUD, DOE, and the Department of Transportation (DOT). The center is a cooperative effort among the seven professional associations representing local governments: the International City Management Association (ICMA), National League of Cities (NLC), National Association of Counties (NACo), U.S. Conference of Mayors (USCM), National Conference of State Legislatures (NCSL), National Governors' Association (NGA), and Council of State Governments (CSG). While the National Community Energy Management Center, administered by the Academy for State and Local Government, has the potential to play a major role in helping local governments achieve energy efficiency in times of dwindling federal and state resources, its own funding is small and its future is uncertain.[15]

Clearly if local governments are to realize the opportunities afforded by CEM, they need information based on assessments of already-existing local energy programs. Communities need information about the barriers faced by local governments in both developing and implementing community energy-management strategies. This chapter is designed to fill that information gap by summarizing the sparse systematic literature on community energy management, emphasizing the findings from the North Carolina Local Energy Policy Project, an ongoing research effort I have conducted since 1979; presenting new data from the most-recent effort of the project; and suggesting both problems and future research areas for studying the design and implementation of community energy-management strategies.

Community Energy Management: Early Research and Theory

The CEM literature falls into five broad categories, listed in order of most to least extensive: background readings; community plans and studies; technical methods, procedures, and models for developing programs; bibliographies; and systematic research, including empirical research, detailed case studies, and other evaluations. Clearly any local government interested in community energy management can turn to hundreds of materials produced in the first four categories but will find very little material in the last category.

Background materials that detail the nature of this nation's energy problems, effects on local governments, and available opportunities are abundant.[16] Similarly, several hundred community energy plans, feasibility studies, and other useful products have been produced by or for local governments at the municipal, county, and area-wide levels that detail existing and planned efforts.[17] Many of these are based on the plethora of technical methods and models developed for CEM by the federal and state governments, consulting firms, and a few innovative local governments.[18] Access to such materials is made simple through the use of a number of comprehensive and annotated bibliographies.[19]

Almost none of the materials in the first four categories, however, were developed after detailed evaluation studies of local programs or after studies developed from the users' perspective. That is, the federal, state, and local governments, along with their professional associations and other interest groups, have proceeded to promote and develop CEM strategies without first examining local readiness for integrating energy into the already-crowded list of local responsibilities. The new National Community Energy Management Center, for example, offers technical assistance, information, and training to local governments but does not have a research component.

Benchmark Studies

The amount of empirical research conducted on CEM is surprisingly lacking, yet initial findings suggest many barriers to CEM that might be addressed through successful research efforts that can identify problems and offer policy guidance. Existing research does offer some benchmark information and the beginnings of a theoretical foundation for studying community energy management.

A framework for examining the energy-related roles of local government can be derived from the five areas cited by Henderson as potential energy-related roles for local government. These are the roles of consumer, producer, regulator, policymaker, and planner, although not every local government performs every role and many responsibilities are shared with other levels of government and the nonprofit sector.[20] Such roles define both an internal and external dimension.[21] Local governments can pursue energy-efficient strategies in their internal operations to save money, and even greater savings can result from external programs that affect such major end uses of energy as space heating and cooling, transportation, and land use.

A growing literature has documented the progress of local governments in responding to the various opportunities for energy management on both the internal and external dimension. This literature has also helped sort out some of the institutional and social barriers to program development and implementation. Studies by various researchers, including a national survey by the International City Management Association and single state studies, show that significant energy activities in terms of financial and/or energy savings have been seriously explored by only a few hundred U.S. cities and counties. This research found that most governments, regardless of population, had not developed extensive internal or external energy programs except for some low-cost, low-risk, usually technical operations within local government.[22] More important, my own separate studies of counties, small governments, and the largest cities in North Carolina found that most communities did not plan on any additional energy activities.[23] Weatherizing public buildings and improving vehicle-fleet maintenance are often addressed, but policies affecting life-style on the internal dimension, such as restricting employee travel or changing employee conservation behavior, are not popular. Similarly, steps to affect community-wide activities—those on the external dimension —are usually limited to citizen-awareness efforts. There has been little effort at incorporating energy efficiency into land-use or zoning policies, for example.

My North Carolina findings correlate to studies by the General Accounting Office[24] and an assessment of energy-efficient land use by the American Planning Association.[25] The latter state-of-the-art survey of 1,426 local, regional, and state planning agencies found only thirteen American communities that had enacted land-use and development regulations explicitly to accomplish energy-savings goals. These included ordinances for heating and cooling, automobile transportation, construction materials and processes, and alternative energy.

On the other hand, follow-up surveys by me[26] in North Carolina and the California Office of Appropriate Technology in that state[27] both showed increases in activities by communities in each state by 1981, suggesting that CEM had captured more local officials' attention after 1979–1980 and before the Reagan administration cutbacks took effect. Still, a nuts-and-bolts approach to energy problems dominated, with uncontroversial internal-operations efforts most likely to occur and activities affecting attitudes or life-style changes rarely achieved.

Several research efforts have looked more closely at the local government role as policymaker, administrator, and planner, covering a range of obstacles to organizing and implementing energy programs. Bowman and Franke examined the attitudinal determinants of support for municipal conservation by elected officials in Texas cities, thus assessing the local government policy-making role.[28] They found local elected officials to be especially crucial actors in the energy-policy process since activities requiring relatively marginal fiscal investments were not promoted due to their implications on the life-styles of community residents. That is, in-house operations programs and citizen-awareness activities are popular with elected officials, but there is much opposition to regulatory programs, especially those affecting commercial and developer elites.

Several analysts, noting that the most-active governments on energy matters have been a few western cities such as Portland, Seattle, San Diego, and Davis, California, along with St. Paul, Minnesota, have examined the CEM histories of each community and found several preconditions for successful community energy management: early and continued participation by elected officials, realization by community leaders and citizens that their economic self-interest suggests an energy strategy, open government, and widespread citizen participation.[29] Detailed case studies of sixteen communities participating in DOE's Comprehensive Community Energy Management Program (CCEMP) also found elected official support to be very important, along with several other key conditions: an early focus on program implementation and selected activities, skillful program management, and emphasis on building on existing programs or activities supporting agreed-upon community objectives.[30]

My separate studies of 59 percent of the counties, 72 percent of the municipalities with populations greater than 10,000 and 119 smaller governments in North Carolina have concentrated on the administrative role of local government in CEM. Key findings suggest that a perceived lack of public support has served to inhibit energy innovation by local officials. Thirty-four percent of the county officials, 39 percent of the medium- to large-city respondents, and 42 percent of the officials from small towns mentioned public support as their "most pressing need" for developing and implementing energy programs. This need received the highest ranking from local officials and was perceived to be even more important than technical assistance.[31]

Research findings on the low levels of existing and planned local energy programs, the need for local elected official leadership, and the perceptions by local officials that public support is lacking are further understood by looking at parallel research findings on individual citizen conservation efforts and a recent survey of citizen support for selected local energy activities.

Numerous studies on individual citizen conservation efforts find convenient, low-cost conservation activities most popular and life-style or attitudinal-changing programs least likely to occur.[32] Three recent reviews of the literature on public opinion and energy clearly document that major conservation steps have not been taken by individuals, nor is the pace of activity dramatically increasing. Perhaps even more important, in light of the fact that the most-significant energy and cost savings can come from external programs dealing with heating and cooling, automobile transportation, land use, zoning, building construction, and other factors affecting life-style, is that socio-economic status is a major determinant of conservation actions, with education, occupation, and age also closely related to conservation actions.[33] It is not surprising that local officials have not felt the push of public demand for local energy activities. Recent public-opinion research in one medium-sized California city, relating to governmental regulatory measures designed to lessen energy consumption in land use, buildings, and other external factors, supports these findings. An important relationship between high levels of affluence and opposition to specific energy conservation actions was found by Neiman and Burt in their 1981 study.[34] This suggests opposition to local government community energy-management plans because the most-affluent residents are also most likely to participate more frequently and effectively to oppose CEM plans. On the other hand, there are hundreds of cities in which citizens have played an organized role in the production and use of energy, as documented in the existing CEM plans. Clearly information about elected official and citizen interest in CEM is important for overcoming obstacles to successful program implementation.

Energy Coordinators

The data presented in this section are drawn from the most-recent in a series of my research efforts aimed at improving the local government role in realizing financial and community-level energy programs. The North Carolina Local

Energy Policy Project began in 1979 with seed money from the Faculty Professional Development Fund of North Carolina State University. Various project activities have addressed the informational, organizational, and other needs of municipal and county governments in the state; developed detailed information and evaluations of local programs; and recommended workable assistance strategies to other levels of government.

Working closely with the municipal and county professional associations in the state, as well as officials at all levels of government, several survey projects have been completed for the Local Energy Policy Project: empirical studies of municipal energy programs, including both large cities and small communities; empirical studies of county-level energy programs; and several follow-up surveys of the role perceptions and informational needs of local energy coordinators.

One recent project activity was a study of the accomplishments, plans, and informational needs of energy coordinators in North Carolina.[35] Mailed questionnaires were sent to the contact persons for energy or the designated energy coordinators for seventy-three communities in late 1980 and early 1981. Key local energy personnel were identified by two fall 1979 surveys completed by the North Carolina League of Municipalities. Those surveys found that energy coordinators-contacts had been designated by all forty-three communities with populations of 10,000 or more and for 150 communities of less than 10,000 population. All of the large communities and thirty of the smaller communities, chosen from a computer-generated table of random numbers, received questionnaires for the study of energy coordinators.

Sixty-nine questionnaires were returned, a response rate of 95 percent. Four of those were returned with acknowledgment that the designated respondent was no longer employed by the community, and one questionnaire was returned several months after data analysis had been completed. Thus, sixty-four questionnaires, 88 percent of the original sample, were used for analysis. This includes 81 percent (thirty-five of forty-three) of the communities with a population of 10,000 or greater in the state and twenty-nine (approximately 20 percent) of the smaller communities identified as having designated energy personnel. Overall, 55 percent of the respondents represent communities with populations of 10,000 or more, and 45 percent represent smaller communities.

Energy contacts-coordinators were asked a series of questions covering their community's organization for energy programs, their own perceptions of energy options available to the community, and the sources of information they used in performing their role.

Following the initial survey, a telephone survey was made in late 1981 to the energy coordinators in the twelve communities identified in the initial survey as being the most active in the state in comprehensive energy management. Respondents were asked a series of questions aimed at assessing in more detail their perceptions of the barriers to comprehensive energy management. This information was combined with a reading of all energy documents, such as plans, ordinances, and news releases, made by each of the communities since 1977.

This section presents data relating to respondent perceptions of the role of elected officials and citizens in their community's energy efforts. Because the data are drawn from communities that have demonstrated an interest in CEM by appointing energy coordinators, perceptions of program development and implementation are based on actual experience with energy programs. Such information should prove particularly useful to communities that have not attempted organized energy efforts.

Background of Community Organization for Energy Programs

Whether from large or small communities, the energy coordinators are amateur energy experts. Only three hold their positions full time, and only four consider themselves to be specialists on local energy matters. Two-thirds of the respondents devote less than 5 percent of their working time to energy matters, and only 7 percent claim to have had their other work responsibilities altered so that they could take on activities within the energy area. Finally, other than the three full-time coordinators in the study, not one respondent claimed to have received a salary increase for bearing the energy-organizing responsibilities for the community.

Ninety-two percent of the respondents were appointed to their positions as energy personnel. Most likely to appoint the energy person was the city manager (52 percent), followed by self-appointment (19 percent), mayoral appointment (16 percent), and a wide variety of other selection processes.

Important to comprehending the role of these energy coordinators-contacts is an understanding of their normal job-related responsibilities. Twenty-four of the respondents (38 percent) are city managers, fifteen (23 percent) are engineers or public-works officials, ten (15 percent) are clerks, and the others hold such positions as finance officer (N = 5, 8 percent) planner (N = 3, 5 percent), or other more-specialized positions. Three are full-time energy coordinators. Without additional authority or resources, these individuals are likely to be hampered in what they can accomplish regarding the energy potential for each community.

The survey of sixty-four energy coordinators updated information collected earlier on the status of energy programs in the state.[36] Low-cost, low-risk, usually technical operations within local government continued to dominate. By mid-1981, only thirteen (20 percent) of the communities had energy budgets for government departments. Only five communities (8 percent) had citizen energy advisory committees, and only sixteen (25 percent) had quantified alternative energy efforts. Fifty-five percent of the communities (N = 35), however, had developed guidelines for energy conservation, and 31 percent (N = 20) had attempted to measure their energy conservation savings. Clearly the communities had attempted more program development since the earlier surveys. (Because such community characteristics as population size, community wealth, and other traditional socioeconomic indicators offer such limited help in understanding differences among communities regarding energy activity, such data are not presented here. Instead, data related to the organization for community energy management and energy coordinator perceptions are used.)

The survey included a series of questions that sought respondent perceptions of the key obstacles to program implementation. On the internal dimension, employee attitudes were cited most frequently by respondents who had attempted internal programs, mentioned first by fifteen (23 percent) of the energy coordinators. Of the nine respondents with experience with external or community-wide energy programs, two-thirds (N = 6) cited citizen opposition first and one-third (N = 3) mentioned lack of interest and expertise by local elected officials as the major obstacle to program implementation.

Survey respondents were most likely to want to pursue only internal energy matters. Two-thirds (N = 36) of the energy coordinators cited only internal concerns as part of their responsibilities, and only twelve (19 percent) of those responding to a direct question about what they considered to be their responsibilities perceived of both an internal and an external energy role. Without exception, respondents who preferred only an internal role cited perceived difficulties with citizen and elected official opposition as primary reasons for role choice.

Despite the overwhelming concern with internal and largely technical energy programs, respondents were most likely to mention leadership and administrative skills as most important to success in their energy roles. Specifically, twenty-five (39 percent) of the respondents mentioned leadership and administrative skills first, while only ten (16 percent) mentioned engineering or technical skills first. These responses indicate frustration with a job that requires achieving consensus—whether on the internal dimension in dealing with other employees and/or elected officials or on the external dimension in dealing with citizens, interest groups, or the elected officials who represent them.

Another set of questions related to coordinator communication with elected officials and citizens. The energy coordinators in this study had minimal communication with either elected officials or citizens regarding energy matters. Thirty-six of the respondents (56 percent) communicated with the mayor of their community on energy matters very rarely or never. Only three (5 percent) discussed energy concerns with the mayor once a week. Similarly,

52 percent of the respondents (N = 33) very rarely or never discussed energy activities with council members. Only six respondents (9 percent) conferred with council members several times per month each. Finally, nearly three-fourths of the respondents (N = 48, 74 percent) never or only very rarely communicated with members of the general public on energy matters. Only two energy coordinators (3 percent) claimed to meet with citizens or citizen groups regularly.

Characteristics of Energy-Active Communities

Communities were scored on an energy-activity scale by summing the number of positive responses to a series of questions about the existence of specific energy-related management actions. The actions spanned the gamut of internal and external energy concerns, dealing both with highly technical and lifestyle issues.

Two major characteristics of energy-active communities emerge from the initial survey of sixty-four communities having designated energy coordinators. Communities having city managers as energy coordinators (N = 24) had significantly more energy activities than communities with a person in another position serving as energy coordinator (X^2 = 11.94, p = 0.036). This is important, especially in light of the fact that three communities have full-time energy coordinators. Given the overwhelming perception by respondents that leadership and administrative skills are most important to the energy role, however, it is not surprising to find the most-active communities having their lead energy person in a generalist role that embodies a great deal of responsibility and opportunity to work for consensus on issues.

Coordinators from the most energy-active communities were no more likely than coordinators from low-energy-active communities to communicate with the city council (X^2 = 2.63, p = 0.621) or with the mayor (X^2 = 5.51, p = 0.702). However, coordinators in the most-energy-active communities had significantly more-frequent communications with citizen groups than coordinators in communities that were not energy active (X^2 = 22.58, p = 0.004).

The organizational variables that best characterize energy-active communities, then, are the types of individuals placed in the position of energy coordinator-contact and the frequency of communication with citizen groups. The first suggests that the policy generalist, a city manager, has both the authority and the ability to achieve consensus among competing ideas and serve in a truly coordinating role. The latter highlights the importance of public support for community energy programs.

To gather more information about the characteristics of energy-active communities, the twelve cities that ranked highest in number and scope of energy program development were surveyed by telephone after the initial survey data were analyzed. Energy coordinators-contacts in each community were asked a series of questions regarding their community's organization for energy management. Community records and other documents relating to the energy role were also examined to aid in interpretation.

Several interesting conclusions can be drawn from the twelve informal interviews. First, each of the communities housed the energy function in a lead office, such as the mayor's or city manager's office, or in a full-time energy coordinator's office. Second, ten of the twelve individuals interviewed claimed that the key to success in their operations was the ability to lead others to achieving policy consensus and commitment.

A third key theme emerging from the telephone interviews was the necessity for coordinators to use an incremental approach for gaining their energy goals. That is, instead of closely abiding by a planned strategy for program development and implementation, these successful coordinators found the most success in an incremental approach. They looked for opportunities, based on their assessments of the political acceptance of ideas, and pursued energy programs that had the greatest chances for success.

Finally, only two of the most energy-active communities had formal citizen advisory committees; however, all of the coordinators stressed their use of citizen information, generally through informal contacts and the use of citizens with technical expertise in selected energy areas. Eight of the twelve individuals interviewed mentioned strategic and highly visible public announcements of successful programs as a wiser approach to achieving citizen and elected official commitment to energy programs than establishment of formal coordinating committees. In the original mail survey, all twelve of the individuals from the most energy-active communities cited employee attitudes, citizen opposition, or problems with elected officials as the key obstacle in implementing their programs.

Conclusion and Future Research Areas

The field of community energy management is a growing concern of research analysts at a particularly important stage in its development. The challenge for local governments is to achieve energy and cost savings without the policy guidance that was available before recent dramatic cutbacks in conservation-program budgets at all levels. The first round of research suggests the need for additional study of several key areas, especially the relationships among elected official attitudes, public support, and the actions of energy administrators. The data presented in this chapter offer the conclusion that elected officials and local citizens play important roles in community energy management, serving often as real or perceived barriers to program development and implementation by energy personnel. The most energy-active communities in the recent North Carolina studies have found ways to overcome such barriers, generally through skillful leadership and management techniques. Comprehensive goals have sometimes been sacrificed to gain politically feasible options. Additional research could enable a larger number of energy personnel to realize the barriers to successful program implementation, a first step in designing wise implementation strategies. We know now that local energy policy is often more a political and management program than the technical jargon usually surrounding it implies.

Notes

1. Steve Hudson, "An Assessment of Local Government Energy Management Activities," Urban Data Service Report 12 (Washington, D.C.: International City Management Association, August 1980).

2. See David Morris, *Self-Reliant Cities* (Washington, D.C.: Institute for Local Self-Reliance, 1982), pp. 129, 130, 189. Also see Annette Woolson, *The County Energy Production Handbook* (Washington, D.C.: National Association of Counties Research Foundation, 1981).

3. For the early history of local involvement in energy management, see Beverly A. Cigler, "Directions in Local Energy Policy and Management," *Urban Interest* 2 (Fall 1980): 34–42.

4. For descriptions of these projects, see Beverly A. Cigler, "Organizing for Local Energy Management: Early Lessons," *Public Administrative Review* 41 (July–August 1981): 470–479, and Beverly A. Cigler, "Intergovernmental Roles in Local Energy Conservation: A Research Frontier," *Policy Studies Review* 1 (May 1982): 761–776.

5. Marion L. Hemphill, "Energy Management from a Community Perspective," *Urban Interest* 3 (Fall 1981): 66–75.

6. Beverly A. Cigler, "Community Energy Planning and Management," *Journal of the American Planning Association* 48 (Spring 1982): 245–248, and Cigler, "Directions in Local Energy Policy and Management," (Fall 1980): pp. 34–42.

7. For a good review of programs, see Henry Lee, "The Role of Local Governments in Promoting Energy Efficiency," in Jack M. Hollander (ed.) *Annual Review of Energy* (Palo Alto, Calif.: Annual Review, 1981), 6:309–337.

8. Cigler, "Community Energy Planning," pp. 245–248.

9. Good descriptive information on these programs may be found in: Morris, *Self-Reliant Cities,* especially the listing of forty-three cities with community energy plans, p. 232; James Ridgeway, *Energy Efficient Community Planning: A Guide to Saving Energy and Producing Power at the Local Level*

(Emmaus, Pa.: J.C. Press, 1980); and Jon van Til, *Living with Energy Shortfall* (Boulder, Colo.: Westview Press, 1982).

10. The major studies on the role of conservation are: Daniel Yergin and Martin Hillenbrand, eds., *Global Insecurity: A Strategy for Energy and Economic Renewal* (Boston: Houghton Mifflin, 1982); Solar Energy Research Institute, *A New Prosperity: Building A Sustainable Future* (Andover, Mass.: Brick House Publishing, 1981); Marc Ross and Robert Williams, *Our Energy: Regaining Control* (New York: McGraw-Hill, 1981); John H. Gibbons and William U. Chandler, *Energy—The Conservation Revolution* (New York: Plenum Press, 1981); Roger Sant, *The Least Cost Energy Strategy* (Pittsburgh: Carnegie-Mellon University Press, 1979); Henry Kendall and Steven Nadis, eds., *Energy Strategies: Toward a Solar Future* (Cambridge, Mass.: Ballinger, 1980); Hans H. Landsberg, chairman, et al., *Energy: The Next Twenty Years,* report by a study group sponsored by the Ford Foundation and administered by Resources for the Future (Cambridge, Mass.: Ballinger, 1979); National Research Council, *Energy in Transition: 1985-2010,* Final Report of the Committee on Nuclear and Alternative Energy Systems, National Academy of Sciences (San Francisco: W.H. Freeman and Company, 1979); and Robert Stobaugh and Daniel Yergin, eds., *Energy Future,* Report of the Energy Project at the Harvard Business School (New York: Random House, 1979).

11. Representative of these are: Conference/Alternative State and Local Policies, *Local Alternative Energy Futures: Developing Economies/Building Communities* (Washington, D.C.: Conference/Alternative State and Local Policies, 1980); National League of Cities, *Financing Local Energy Programs* (Washington, D.C.: National League of Cities and National Community Energy Management Center, 1981); and items listed in notes 16–20 of this chapter.

12. See Ross and Williams, *Our Energy,* for a comprehensive analysis of a market-oriented energy conservation strategy and its drawbacks.

13. Cigler, "Intergovernmental Roles," pp. 761–766.

14. For example, see John L. Moore, *The Comprehensive Community Management Program: An Evaluation* (Argonne, Ill.: Argonne National Laboratory, December 1981).

15. Beverly A. Cigler, "Energy Center Assists Local Officials," *Public Administration Times,* May 1, 1982, p. 4.

16. See n. 10 for the best of this literature. Also see the following important works: Robert W. Burchell and David Listokin, eds., *Energy and Land Use* (Piscataway, N.J.: Center for Urban Policy Research, Rutgers, 1982); Joel T. Werth, ed., "Energy in the Cities Symposium," *Planning Advisory Service Report No. 349* (Chicago: American Planning Association April 1980); and Steve Hudson, "Managing the Impact of the Energy Crisis: The Role of the Local Government," *Management Information Service Report* 12 (Washington, D.C.: International City Management Association, February 1980).

17. Representative examples are: City of St. Paul, *The St. Paul Experience,* 5 vols. (St. Paul, Minn.: Department of Planning and Economic Development, 1980); City of Baltimore, *Energy in Baltimore: A Commitment to the Future* (Baltimore: Mayor's Office of Energy, 1981); *and Franklin County Energy Project, Franklin County Energy Study: A Renewable Energy Future* (Amherst, Mass.: University of Massachusetts Press, 1979).

18. The most-prominent examples are: Hittman Associates, *Comprehensive Community Energy Planning,* 3 vols. (Washington, D.C.: U.S. Department of Energy, 1978); Alan Okagaki and Jim Benson, *County Energy Plan Guidebook: Creating A Renewable Energy Future* (Fairfax, Va.: Institute for Ecological Policies, 1979); Sizemore and Associates, *Methodology for Energy Management Plans for Small Communities* (Washington, D.C.: U.S. Department of Energy, 1978); Eileen Baumgardner and Don Schultz, *Local Energy Planning Handbook* (Sacramento, Calif.: California Energy Commission, 1981); and T. Owen Carroll, *The Planner's Energy Workbook: A Manual for Exploring Relationships between Land Use and Energy Utilization* (Washington, D.C.: U.S. Department of Energy, 1977), by Brookhaven National Laboratory and State University of New York at Stony Brook.

19. Examples are: HUD User, *Energy Conservation for State and Local Government* (Germantown, Md.: HUD User, 1980) and U.S. Department of Energy, "Guidebook Resources for Local Governments" (Washington, D.C.: U.S. Department of Energy, April 1981).

20. Lenneal Henderson, "Energy Policy and Urban Fiscal Management," *Public Administration Review* 41 (January 1981): 158–164.

21. Cigler, "Directions in Local Energy Policy and Management," pp. 34–42, and Cigler, "Organizing for Local Energy Managements," pp. 470–479.

22. Hudson, "An Assessment of Local Government Energy Management Activities"; James L. Franke and Ann Bowman, "The Development of Community Energy Conservation Policy: Determinants of Individual Decision Making" (Paper presented at the Annual Meeting of the Southern Political Science Association, Memphis, Tennessee, November 5–7, 1981); James L. Franke and Ann Bowman, "Energy Conservation in a Pro-Consumption Environment" (Paper presented at the Annual Meeting of the Southern Political Science Association, Atlanta, Georgia, November 6–8, 1980); Cigler, "Organizing for Local Energy Management," pp. 470–479.

23. See n. 22 above and Beverly A. Cigler and J. Kent Crawford, "Energy Problems: Untapped Cost Savings for Small Governments," *Municipal Management: A Journal* 5 (Summer 1982): 39–47. as well as Cigler, "Intergovernmental Roles," pp. 761–776.

24. U.S. General Accounting Office, *Greater Energy Efficiency Can Be Achieved through Lane Use Management,* Report to the Congress EMD-82-1 (December 21, 1981).

25. Duncan Erley and David Mosena, "Energy-conserving Development

Regulations: Current Practice," *Planning Advisory Service Report No. 352* (Chicago: American Planning Association, August 1980).

26. Some findings from this data set are found in Beverly A. Cigler, "Implementing Local Programs: The Place of Position" (draft manuscript, August 1982).

27. California Office of Appropriate Technology, *Local Energy Initiatives: A Second Look. A Survey of Cities and Counties* (Sacramento: California Office of Appropriate Technology, December 1981).

28. Franke and Bowman, "Development of Community Energy Conservation Policy."

29. Henry Lee, "The Role of Local Governments in Promoting Energy Efficiency," in Hollander, Simmons, and Wood, eds., *Annual Review of Energy,* 6:309-337; and Cigler, "Intergovernmental Roles," pp. 761-776.

30. Moore et al., *Comprehensive Community Energy Management Programs.*

31. Cigler and Crawford, "Energy Programs"; Cigler, "Organizing for Local Energy Management," pp. 470-479; and Beverly A. Cigler, "Organizing County Energy Programs: Early Lessons" (unpublished preliminary draft, 1981).

32. William H. Cunningham and Sally C. Lopreato, *Energy Use and Conservation Incentives: A Study of the Southwestern United States* (New York: Praeger, 1977), and Paul P. Craig, Joel Darmstadter, and Stephen Rattien, "Social and Institutional Factors in Energy Conservation," in Jack M. Hollander (ed.) *Annual Review of Energy* (Palo Alto, Calif.: Annual Reviews), 1: 535-551.

33. See van Til, *Living with Energy Shortfall;* Marvin E. Olsen, "Public Acceptance of Energy Conservation," in Seymour Warkov, ed., *Energy Policy in the United States; Social and Behavioral Dimensions* (New York: Praeger, 1978); and Barbara C. Farhar, "Public Opinion About Energy," in Jack M. Hollander, Melvin K. Simmons, and David O. Wood, eds., *Annual Review of Energy,* (Palo Alto, Calif.: Annual Reviews, 1980), 5:141-172.

34. See chapter 8.

35. Some findings from this data set are found in Beverly A. Cigler, "Implementing Local Programs: The Place of Position" (draft manuscript, August 1982).

36. See Cigler, "Organizing for Local Energy Management," pp. 470-479, and Cigler and Crawford, "Energy Programs."

8

The Political Constraints on Energy-Conservation Policy: The Case of Local Citizen Receptivity

Max Neiman and
Barbara J. Burt

This chapter deals with an analysis of resident support for a set of proposed policies designed to raise the level of energy conservation in one city. We begin with the assumption that the conservation of energy is and will continue to be an important component of energy policy. Despite the apparently low priority given by the current national administration to government encouragement of conservation, market pressures and the actions of states and localities will ensure that conservation will remain important in the overall struggle to manage the energy crisis. Although there are substantial variations among the many studies estimating potential energy savings due to conservation, there seems to be agreement that such savings can be quite significant (Landsburg 1979; National Research Council 1979; Stobaugh and Yergin 1979). For example, in examining the electricity needs of California, the California Energy Commission (1981) concluded that increased energy efficiency through conservation could save as much energy as is produced by eight large nuclear or coal-fired plants. The purpose of this chapter, however, is not to dwell on the efficacy of conservation strategies but to focus on some selected issues that are likely to accompany an effort by local governments to contribute to government-stimulated conservation efforts.

As Cigler (1982, pp. 762–763) has pointed out, there are several important reasons for considering the role of local government in managing energy through conservation. First, local governments are in a better position to tailor energy policies to their unique climatic, economic, and demographic characteristics. Next, with regard to conservation, localities may realize substantial benefits through conservation, not merely through the direct energy savings but through the multiplier effects that might be associated with greater disposable income available to the community. Even when the costs of conservation are similar to that of purchasing additional supplies from outside the locale, a greater proportion of the conservation spending is in the local or regional economy. This contrasts with the exporting of local dollars out of the region when energy is purchased from a wholesaler that invests energy expenditures in some other locale or region. Additionally, local government police-power authority can be directed at energy-conservation objectives. Regulations regarding construction standards, land use, and weatherization have been the subject of considerable discussion as they might apply to the achievement of energy efficiency (Comptroller General of the United States 1981). Finally, there are those for whom local government is seen as an important base for countervailing the tendency of society to be organized around large-scale systems of social control (Orr 1979).

Despite the presumed stake that local governments have in adopting an energy-conservation program, studies suggest that very little general advance has been made by localities in adopting energy-conservation policies, although notable exceptions tend to be trumpeted in repetitive fashion. On the basis of considerable research and reviews of national, state, and local materials, Cigler concludes that local governments tend to have modest energy-conservation programs, which tend to focus on in-house improvements in energy use, such as improving the municipal auto fleet mileage or encouraging car pooling among public employees. Cigler (1982, p. 768) identifies a number of factors that have militated against an increased local government role in general conservation, including the absence of national leadership or coordinated intergovernmental action, lack of public awareness of or demand for conservation, shifting burdens of energy costs (as energy prices stabilize, public awareness and concern evaporate), fiscal stress among localities, which prevents investment in energy-policy innovation, a dearth of energy expertise at the local level, and a lack of information sharing about viable programs among local governments or between levels of government.

These factors are undoubtedly important in accounting for much of the paucity and modesty of local energy-conservation programs. There are also a variety of political obstacles that are unlikely to disappear even with improved levels of resources or expertise for developing energy-conservation programs. It must be noted that energy-performance standards for new construction are likely to have mixed receptions in the business, especially development, community. For example, in California, new performance standards for residential construction were to have taken effect in the summer of 1982. Because of the current malaise in the residential construction industry and general hostility of builders to government regulations, there was intense opposition to these new standards. The result is that the new state-imposed energy-performance standards are being delayed until 1983, when the fight will emerge once more. Residential audits or other regulations regarding weatherization and design will perhaps be seen differently by residents depending on life cycle, size of household, income, and the like. Moreover, aside from the expense of regulatory approaches to conservation, for the implementing agency as well as the target of regulation, there are such issues as political conservatism and its suspicion of government meddling in private matters. In short, an energy-conservation program is likely to be a political as well as a technical challenge.

It is the purpose of this chapter to examine some of the potential obstacles to a particular set of conservation efforts under consideration in one medium-sized city in southern California.

Setting and Data Gathering

The data for this study are based on a telephone survey conducted in a southern California city with a 1980 estimated population of 171,000. The city has been incorporated for nearly a century and has a diversity of residents and a mixed economic base. Moreover, the city has a municipal utility that distributes electricity and that has several investments in conventional and unconventional electricity supplies, including geothermal and nuclear energy. The city has been the scene of considerable discussion of energy policy, resulting not only in public debate about the appropriateness of particular energy investments but also stimulating major innovations in government structure. The city has created a citizen's advisory committee (the city Energy Commission) and a division within the Utility Department to coordinate energy activities designed to improve energy efficiency. Each of these activities has been well publicized and occasionally controversial.

The city's electric utility is governed by the city council but is directed by a citizens' board, appointed by the council. The tendency for energy policy to be politicized has become especially great in recent years, not only because of citizen unrest regarding energy costs but also because of the involvement of groups interested in soft paths and energy conservation and who believe that previous policy has been overly dominated by traditional supply orientations. In short, the policy area of energy is not a matter of arcane and narrow interest in this city. The presence of the city utility, the existence of several official citizen bodies, and the presence of insurgent consumer groups as well as environmental and antinuclear activists have diffused and publicized the variety of energy issues throughout the community.

An interview schedule for this study was developed after relevant, previous research had been thoroughly reviewed (for example, Honnold and Nelson 1979; Pierce 1979; Sears et al. 1978; Anderson and Lipsy 1978; Buttell and Flinn 1978; Dunlap 1975). On the basis of the review and pretest and keeping in mind necessary limitations on the interview length, questions were designed to measure variables previously found to be important in explaining support for or opposition to government involvement in environmental protection and resource conservation. Respondents were asked about their views of the energy problem, their level of support for particular energy-related policies, their resistance or support concerning government intervention in the economy, and their socioeconomic background. In administering the telephone interviews, 471 heads of household or their respective spouses were surveyed.

The respondents for the study were identified by means of the random-digit dialing, plus-one algorithm (Landon and Banks 1977; Tyebjee 1979). The sample-size goal was originally 500, which allows parameter estimates of 95 percent probability within a narrow confidence interval (Blalock 1972). To ensure that the sample was representative of the general population, it was compared to selected findings of the 1978 Special Census. Table 8–1 summarizes the comparison and supports the assumption that this study's sample is reasonably representative. In sum, the data provide an opportunity to explore citizen support for a local government's proposed set of policies to promote energy efficiency and inspire the conservation of energy.

In this study, the dependent variable is an index of support for city promotion of energy conservation. The index was constructed by a summed rating of the following seven statements:

1. The city of Riverside should provide low-cost loans for residents to improve their homes' energy efficiency.
2. Information about utility costs for residents should be legally required at the time of sale of a house or with a new renter.
3. It would be wrong for the city to require residences to have a minimum level of energy efficiency.
4. The city should require that owners add such things as insulation, caulking, weather stripping, or attic fans to improve energy efficiency.
5. When a house is sold, the city should require an energy audit of the residence to determine the energy efficiency of the residence.
6. The city should regulate building and landscaping so that access to sunlight is maintained for solar energy.
7. The city should spend money to inform people on ways to increase the efficiency of energy use.

The respondents could answer "strongly agree," "agree;" "undecided," "disagree," or "strongly disagree" to the seven statements. Agreement with statement 3 actually denotes opposition to city energy policy so its score was subtracted from the total scored on the other six statements. Recognizing that it was possible in this procedure for some of the statements to be superfluous, t-tests were used to evaluate whether a statement distinguished between the opponents and supporters. This test led us to eliminate statements 1, 2, and 7. Thus, scores on the four remaining statements ranged from 1 to 16 (Edwards 1957).

Based on the distribution of the scores, the respondents were divided into three groups: supporters, opponents, and the undecided. The low scores, 1 to 6, denote agreement with city promotion of energy-conservation policy, and these respondents were labeled supporters (N = 132). The highest scores, 10 to 16, denote disagreement with a role for the city in implementing energy policy, and these respondents became the opponents (N = 133). Finally, the middle range of scores, 7 to 9, was assigned to an undetermined group (N = 121). The mean index scores are 4.7, 11.7, and 8.0 for the supporters, opponents, and undetermined groups, respectively.

The proposed policies that comprise the policy index are not exhaustive. A variety of other policies might have been included. The focus on this set resulted from their being actually considered and discussed within the study site. Since considerable public discussion over these policies had occurred, it was felt that they should be the locus of concern.

In specifying relevant variables and their measures for support for energy conservation, there were no available theories to guide selection. Research in several similar areas, however, does provide some instruction. For example, recent studies of environmental protection have suggested a nexus between support for the conservation of resources and beliefs and attitudes regarding environmental protection. Pierce (1979) provides some evidence for the idea that environmentalism and support for resource conservation are positively related, suggesting that the variables correlated with environmental support might be similarly related to energy conservation. In this connection, one might expect measures of economic status and liberal political attitudes to be positively related to support for government efforts to promote energy conservation (Van Liere and Dunlap 1980; Honnold and Nelson 1979).

Analyses of household energy consumption are also suggestive regarding correlates of support for energy conservation (Burby and Bell 1978; Unseld 1979). This research reveals significant connections between energy consumption and size of residence, income, number of household members, and whether the residence is rented or owned. It is not surprising to find such relationships; the more-subtle implication might be that attributes of a household pose a matrix of self-interest and rationality, with more-affluent residents in larger homes, households with many members, and home owners generally being disposed against mandated energy conservation, since they tend to consume more energy. So while studies of environmental protection suggest

Table 8-1
Comparison of Selected Special Census and Study Measures

	City Special Census	Study Sample
Percent over 65	8.5	9.5
Percent Black	6.6	4.9
Percent Hispanic	10.2	7.4
Owner occupied	69.7	71.8
Renter	27.1	26.8
Median family income	$14,424 (1977)	$23,000 (in 1981 dollars) $14,000 (in 1977 dollars)

positive relationships between socioeconomic status and support for energy conservation, it is also possible that such people might find their narrower definitions of self-interest in conflict with such policies. On the other hand, the less affluent, since they consume less energy, might support conservation. But what if such policies are perceived to increase housing costs? Because structural improvements result in immediate costs, while the energy savings might take years to compensate for the initial costs, renters might oppose such policies if they see their stay at a residence as temporary and short-term.

Finally, many energy-conservation proposals, including a number of those studied here, imply considerable extensions of government regulation into property rights, development, and life-styles. Consequently it is reasonable to propose that measures of political attitudes, such as general liberal-conservative inclinations and views concerning government intervention in the economy, might also be related to support for government policies promoting energy conservation.

In considering the extant research, then, the following measures of those variables posited as predictors of support for government-mandated energy conservation are used in this study: (1) type of residence, (2) age, (3) years of schooling, (4) total family income, (5) ethnic identification, (6) number of rooms in the residence, (7) number of individuals in the household, (8) partisan identification, (9) self-rated political philosophy on a liberal-conservative continuum, (10) summed-scale indicating degree of support for government involvement in the economy, and (11) summed-scale indicating degree of support for environmental protection. The precise operations for these measures are described in the Appendix to this chapter.

Measures 1 through 7 are background measures selected on the basis of their demonstrated importance in other, related studies. Measures 8 through 11 are indicators of attitudes that presumably incline people for or against greater government involvement in the economy. In the case of partisanship, it is proposed that Republicans are less likely to support government-mandated conservation, and Democrats are more likely to support it. Those labeling themselves as politically or economically conservative are hypothesized to be less likely to support city promotion of energy conservation. Finally, supporters of environmental protection are more likely to support government efforts to encourage or require greater energy efficiency. Table 8-2 reports the bivariate intercorrelations among these measures. The highest intercorrelation does not reach an absolute value of 0.50, and there is no worry regarding problems of multicollinearity in the subsequent multivariate analysis.

Analysis

The analysis of the selected variables proceeds in two stages. The first is a bivariate analysis between the independent variables and the index of support

Table 8-2
Intercorrelations among Predictors of Support for Energy-Conservation Index (Spearman's Rho)

	Residence Type	Age	Schooling	Income	Ethnic	Number of Rooms	Number of Residents	Partisanship	Political Philosophy	Economically Conservative	Environmental Support
1. Residence type	1.000										
2. Age	.313	1.000									
3. Schooling	.105	-.288	1.000								
4. Income	.433	.051	.311	1.000							
5. Ethnic	.376	.110	.119	.050	1.000						
6. Number of rooms	.396	.086	.216	.440	.071	1.000					
7. Number of residents	.126	-.164	.014	.238	-.148	.185	1.000				
8. Partisanship	.025	-.063	.061	-.007	.214	.026	.034	1.000			
9. Political philosophy	.180	.177	-.031	.025	.087	.007	-.014	.477	1.000		
10. Economically conservative	.137	.164	.129	.092	.191	.113	-.061	.361	.298	1.000	
11. Environmental support	.100	.155	-.111	-.057	.054	.012	.000	.244	.240	.193	1.000

Note: All correlations with absolute value equal to or greater than 0.078 are statistically significant.

for city promotion of energy conservation. In the second stage, a multivariate analysis is conducted so that the collective strength of the relationship among the explanatory variables can be identified. The bivariate analysis is presented in table 8-3. The index of support for energy conservation consists of three groups: the supporters, the opponents, and the undetermined. The data indicate that eight of the eleven hypothesized variables have a significant (0.05 level is the minimum) relationship with the support index.

In this stage of the analysis, four status variables are significantly related to the group of respondents who support city involvement in energy-conservation policy. Renters, minorities, the young, and persons living in smaller residences appear to be supporters. Additionally, four variables, characterized as political attitudes, are related to the support index. Partisanship is the weaker of these four variables, while the environmental-support index demonstrates a relatively strong pattern among supporters and opponents. The economic-conservatism index and self-rated political ideology support the findings of the other two political variables. Clearly those who have less opposition to government intervention in the economy also support local government action in promoting energy-conservation measures.

There was some suspicion that the undetermined group was clouding the results of the bivariate analysis because their responses to the questions making up the support for city policy index tended to be inconsistent. The gamma values associated with each cross-tabulation appear to support our suspicion. When the cross-tabulation was done excluding the undetermined group (see the last column of table 8-3), the gamma value increased for every variable, suggesting that the independent variables are more erratically distributed for the undetermined group than for the supporters or opponents. Because the purpose of the research is to clarify the factors that lead to support or opposition to city-promoted conservation policy, it was concluded that the multivariate analysis would be more definitive if the undetermined group of respondents was excluded.

Still unanswered is the question of which factors are the most important in identifying supporters and opponents and what is the collective contribution of these factors in differentiating supporters and opponents. Discriminant function analysis (Klecka 1980) is a technique suitable for nominal-level variables, and the procedure enables analysis of differences between two groups according to specified variables simultaneously.

Table 8-4 reports the results of the stepwise discriminant function analysis employing Rao's V criterion for the selection of variables.[1] The results are statistically significant, as indicated by the chi-square of the function's final Wilk's lambda value. Additionally, the degree of separation between the two groups, supporters and opponents, is acceptable (eigenvalue = 0.310, canonical correlation = 0.487). Supporting the evidence of discrimination between the groups is the range of separation between the group centroids for the supporters (−0.574) and opponents (0.534).

A final indication that differences do exist between the two groups, based on the independent variables, is the classification procedure of the function. Overall more than 77 percent of the respondents were assigned to the correct group. The function classified 74.2 percent of the supporters of city conservation policy correctly and over 80 percent of the opponents. This percentage of correct classification supports our assumption that socioeconomic status and political ideology affect the attitude a respondent will have toward energy-conservation policy.

Six of the eleven original analysis variables survive the stepwise procedure. In order to assess the relative importance of each of the measures in the discriminant function, the column reporting the standardized function coefficient is consulted (see table 8-4). The findings indicate that three of the socioeconomic background measures are important in differentiating between supporters and opponents of increased city promotion of energy conservation. The respondents who are nonminority, live in larger homes or apartments, and are older tend to be opponents, while their counterparts are supporters. These findings are consistent with the energy-consumption studies that reveal affluent residents consume more energy (Newman and Day 1975). According to the data in table 8-2, the correlation between annual family income and number of rooms in the household is substantial ($r_s = 0.440$). Hence the explanation that argues affluent residents, because of living in larger quarters, consume more energy and are perhaps more likely to see required energy-conservation measures as conflicting with life-styles is supported by these data. The data in table 8-2 further indicate a modest relationship between the ethnicity-race measure and family income (-0.288). Additionally, the data indicate a moderate relationship between ethnicity-race and the type of residence in which the respondent lives ($r_s = 0.376$), with minorities more likely to be renters than nonminorities. Since ethnic and racial minorities tend to be less affluent and renters, the data suggest that their support for government-mandated energy conservation might be rooted in their lower status and different self-interest. Renters may regard lower energy costs as a way of reducing their living expenses.

The data also support the expected relationship between the receptivity to government-mandated energy conservation and the three attitudinal measures that survived the stepwise discriminant function analysis. Indeed the index of support for environmental protection is the second most important variable in terms of discriminating power respecting supporters and opponents of energy conservation, based on the value of the standardized function coefficient (0.364). Moreover, the other political attitudes, self-rated political philosophy and economic conservatism index, reinforce the notion that the issue of local energy conservation mandated by local government is likely to become enmeshed by the larger issue of appropriate government scale. According to these data, then, support is found for the linkage shown elsewhere between attachment to environmental protection and agreement with govern-

Table 8-3
Cross-Tabulation between Index of Conservation Support and Independent Variables

Variables	Supporter	Undeter-mined	Opponent	N	Chi-Square	Signifi-cance Level	Gamma	Gamma without Undeter-mined Group
Type of residence								
Renter	47.4	29.3	23.3	116	14.428	.000	.330	.465
Owner	29.8	30.1	40.1	322				
Age								
Less than 21	20.0	40.0	40.0	5				
21–35	45.9	30.1	24.0	196				
36–50	28.2	32.7	39.1	110				
51–59	28.6	26.8	44.6	56				
More than 60	18.3	28.3	53.3	60	29.255	.000	.310	.439
Years schooling								
Less than high school	27.8	50.0	22.2	36				
Completed high school	37.9	27.4	34.7	95				
Some college	34.4	31.7	33.9	183				
Completed college	28.1	26.6	45.3	64				
Beyond college	36.5	25.4	38.1	63	11.134	.194	.066	.095
Family income								
$10,999 or less	45.5	32.7	21.8	55				
$11,000–20,999	31.9	34.8	33.3	69				
$21,000–30,999	41.4	22.4	36.2	116				
$31,000–40,999	30.0	31.4	38.6	70				
$41,000–50,999	40.0	32.0	28.0	25				
Over $51,000	31.6	31.6	36.8	19	9.593	.477	.084	.128
Race								
White	29.7	30.3	40.1	347				
Other	52.9	32.2	14.9	87	23.508	.000	.460	.654

				N				
Number of rooms								
Less than 2	44.4	22.2	33.3	9				
Less than 3	38.5	46.2	15.4	26				
Less than 4	48.6	31.4	20.0	35				
Less than 5	38.1	33.3	28.6	63				
Less than 6	30.1	35.5	34.4	143				
More than 6	31.3	25.7	43.0	164	18.265	.051	.190	.280
Number of persons								
1–3	34.7	29.2	36.1	277				
4–7	34.4	31.8	33.8	154				
8 or more	42.9	28.6	28.6	7	0.603	.963	-.029	-.042
Partisanship								
Strong Democrat	34.0	28.0	38.0	50				
Leans Democrat	47.6	25.7	26.7	125				
Leans Republican	30.1	30.1	39.8	128				
Strong Republican	23.1	32.1	44.9	78	14.071	.029	.174	.251
Political philosophy								
Very liberal	55.6	33.3	11.1	18				
Liberal	44.2	31.2	24.7	77				
Tends liberal	44.9	25.8	29.2	89				
Tends conservative	27.4	28.2	44.4	117				
Conservative	25.8	37.1	37.1	89				
Very conservative	26.3	10.5	63.2	19	29.071	.001	.242	.353
Economic conservatism								
Low opposition to intervention	47.2	28.9	23.9	142				
Moderate opposition to intervention	34.1	30.4	35.6	135				
High opposition to intervention	23.3	31.5	45.2	146	21.339	.000	.294	.421
Environmental support								
High support	53.8	18.8	27.4	117				
Moderate support	31.9	36.9	26.4	110				
Low support	27.4	31.3	50.0	137	33.326	.000	.323	.429

Table 8-4
Stepwise Discriminant Function Analysis: Two-Group Index of
Energy-Conservation Support

	Rao's V	Change in V	Signifi-cance of Change	Standardized Function Coefficient
Ethnicity/race	20.25	20.25	.000	0.511
Environmental support	33.87	13.62	.000	.364
Number of rooms	43.06	9.18	.002	.356
Economic conservatism	51.07	8.01	.005	.290
Age of respondent	55.54	4.47	.034	.284
Political philosophy	59.73	4.19	.041	.271

Notes: Eigenvalue of function = 0.310; final Wilk's lambda = 0.762; canonical correlation = 0.487; significance of function = 0.000; group centroid values = −0.574 (supporters), 0.534 (opponents). Classification results: classified cases that are correctly classified on the basis of function, 77.3 percent; supporters correctly classified, 74.2 percent; opponents correctly classified, 80.5 percent.

ment-mandated energy conservation. Additionally, the data indicate that if energy conservation is translated into specific police-power regulations by local government, those with an antipathy toward government intervention in previously private matters will also assess energy conservation in terms of some antigovernment theme.

Conclusion

Energy-conservation programs are promoted and accepted by many as providing a valuable contribution to the nation's energy supply (Stobaugh and Yergin 1979). Nonetheless, municipal officials charged with the responsibility of providing a steady and adequate supply of electricity to the community are often skeptical about too much reliance on conservation measures as a means of immediate supply and as a means of meeting future demand for energy. Absolute surety of knowing how much will be conserved and how consistently the energy will be conserved is not possible when thousands of individual consumers must be relied upon to cooperate voluntarily. One way of dealing with this situation is by enacting various ordinances mandating specific programs that affect levels of electricity consumption. For example, requiring increased insulation and other energy-efficiency measures in new construction can be projected to save an average number of kilowatt-hours each year. Local legislators, however, are likely to meet organized opposition from builders, realtors, and lenders because these measures will push up the cost of housing construction.

A countervailing force to this organized opposition could be a consensus of support for energy-conservation programs among community residents. The research reported here is an effort to identify the factors that may be important in building and identifying support for local energy-conservation policy. We have attempted to isolate those sociopolitical factors that may determine who, among community residents, will support or oppose efforts to mandate energy conservation. It is reasonable to suppose that conflict and opposition, if it emerges, will center around those who see themselves differentially burdened by such policies. Pierce (1979) argues that perceived self-interest is an important determinant of support for resource conservation, and the results of this study suggest that supporters and opponents of specific local energy programs may also form their attitudes and choose their issue stance based on beliefs about the effect on their life-styles. For example, opponents of city promotion of energy conservation appear to be the affluent or at least the higher-status residents. Although other studies have often found these people to be the supporters of environmental preservation and resource conservation measures, perceived self-interest, perhaps a change in life-style, tends to counter the desire for the other values. Rather than a direct relationship between status and energy conservation, the relationship may be attenuated by the reality of the impact on the person's daily life.

Related to the concept of self-interest, we found that energy-conservation programs, especially if characterized as government regulation, lead respondents to incorporate ideological inclinations into their evaluations of energy-conservation actions. Economic and political liberals tend to be more supportive of government-mandated conservation measures. The responses to the questions comprising the environmental index support the assumption about the liberal-conservative dimension of respondents' attitudes. Although it is well established that conservation is favorably regarded by a majority of Americans and a signficant number of people report taking steps to reduce their energy consumption, the data in this survey appear to demonstrate that specific measures, entailing elements of coercion and manipulation, generate an uncertainty of opinion regarding those programs. One possible explanation for this uncertainty may be the implication that government will have to be enlarged to implement these programs, resulting in a bureaucratic structure with increasing intrusion into previously private and voluntary activity. As such, local promotion of energy conservation through regulation is likely to be complicated by contemporary concern over the scale and cost of government.

Beyond the immediate implications of the survey data, there are additional political constraints on implementing local energy-conservation programs. For example, in cities like the study site that have a municipal electric utility, more than 2,000 in the United States, the general fund is often dependent on receiving revenue from the sale of energy to pay for other city services.[2]

Consequently both the utility and the city are dependent on the sale of electricity, and the promotion of conservation will cut back on revenues. When this implicit contradiction is made explicit, some adjustment will have to be made to maintain certain levels of revenue. Should consumers be asked to pay for their savings in higher electrical charges or in a surcharge? Surely some in the community will see this as a penalty for acting responsibly in conserving resources.

A second political concern that may arise concerns the role the local utility should play in redistribution of the community's resources. Enactment of a program to provide low-cost loans for the lower-income residents to improve their homes' energy efficiency is likely to raise questions about whether all ratepayers should be required to pay for a program that serves only a special class of electricity consumers.

These examples point to the overriding conclusion of this study: conservation policies among localities are likely to encounter an array of obstacles to which conservation proponents are now becoming more sensitive. Technical and economic information regarding conservation programs must also incorporate the social, political, and ideological factors that appear to affect citizen attitudes about energy programs.

Notes

1. Rao's V is a generalized measure of the distance between group centroids. It is used to obtain a measure of total group separation.

2. Municipal utilities often are required by the city charter or by ordinance to contribute a portion of their revenue to the general fund. In the study site, this amounts to 11.5 percent of the general fund each year.

References

Anderson, Richard W., and Lipsey, Mark W. 1978. "Energy Conservation and Attitudes toward Technology." *Public Opinion Quarterly* 42 (Spring): 17–30.

Blalock, H.M., Jr. 1972. *Social Statistics*. New York: McGraw-Hill.

Burby, Raymond, and Bell, A. Fleming, eds. 1978. *Energy and the Community*. Cambridge, Mass.: Ballinger Publishing Company.

Buttel, F.H., and Flinn, W.L. 1978. "Social Class and Mass Environmental Beliefs: A Reconstruction." *Environment and Behavior* 10 (Spring): 433–450.

California Energy Commission. 1981. *1981 Biennial Report to the Governor and the Legislature*.

Cigler, B.A. 1982. "Intergovernmental Roles in Local Energy Conservation: A Research Frontier." *Policy Studies Review* 1 (N.D.): 761–776.

Comptroller General of the United States. 1981. *Greater Efficiency Can Be Achieved through Land Use Management.* Washington, D.C.: General Accounting Office.

Dunlap, Riley E. 1975. "The Impact of Political Orientation on Environmental Attitudes and Action." *Environment and Behavior* 7 (December): 428–442.

Edwards, Allen L. 1957. *Techniques of Attitude Scale Construction.* New York: Appleton-Century-Crofts.

Honnold, J.A., and Nelson, L.D. 1979. "Support for Resource Conservation: A Prediction Model." *Social Problems* 27 (December): 220–233.

Klecka, W.R. 1980. *Discriminant Analysis.* Beverly Hills, Calif.: Sage Publications.

Landon, E. Laird, and Banks, Sharon K. 1977. "Relative Efficiency and Bias of Plus-One Telephone Sampling." *Journal of Marketing Research* 42 (August): 294–299.

Landsburg, H. 1979. *Energy: The Next Twenty Years.* Cambridge, Mass.: Ballinger Publishing Company.

National Research Council. 1979. *Energy in Transition: 1895–2010.* San Francisco: W.H. Freeman and Company.

Orr, D.W. 1979. "U.S. Energy Policy and the Public Economy of Participation." *Journal of Politics* 41 (November): 1027–1056.

Pierce, J.C. 1979. "Water Resource Preservation, Personal Values, and Public Support." *Environment and Behavior* 11 (June): 147–161.

Sears, D.O. 1978. "Political System Support and Public Response to the Energy Crisis." *American Journal of Political Science* 22 (Fall): 56–82.

Solar Energy Research Institute. 1981. *A New Prosperity: Building a Sustainable Future.* Andover, Mass.: Brick House Publishing.

Stobaugh, Robert, and Yergin, Daniel, eds. 1979. *Energy Future: Report of the Energy Project at the Harvard Business School.* New York: Random House.

Tyebjee, Tyzoon T. 1979. "Telephone Survey Methods: The State of the Art." *Journal of Marketing* 43 (Summer): 68–78.

Unseld, Charles T. 1979. *Sociopolitical Effects of Energy Use and Policy.* Washington, D.C.: National Academy of Sciences.

Van Liere, K.D., and Dunlap, R.E. 1980. "The Social Bases of Environmental Concern: A Review of Hypotheses, Explanations, and Empirical Evidence." *Public Opinion Quarterly* 44 (Summer): 181–197.

Appendix 8A: Description of the Study Variables

The following is a detailed description of the study measures. Each background variable is listed with the response categories.

1. Type of residence: Rental scored as 0 and home owners scored as 1.
2. Age of respondent: Younger than 21; 21–35; 36–50; 50–59; more than 60.
3. Year of schooling: Less than high school; completed high school; some college; completed college; beyond college.
4. Annual family income: Less than 10,999; $11,000–20,999; $21,000–30,999; $31,000–40,999; $41,000–50,999; $51,000 or more.
5. Ethnic-Racial: Ethnic-race scored as 0 and nonethnic-race scored as 1.
6. Number of rooms in the household: 2 or fewer; 3; 4; 5; 6; more than 6.
7. Number of persons in the household: 1–3 persons; 4–7 persons; more than 7 persons.
8. Partisian identification: Respondents were asked, "Do you consider yourself a strong Democrat/Republican, or do you only lean Democrat/Republican?"
9. Self-rated political philosophy: Respondents were asked, "If you had to rate your political philosophy, which of the following would you say applies to you: very liberal, liberal, tend to be liberal, tend to be conservative, conservative, or very conservative?"
10. Attitudes toward government intervention in the economy: Each respondent was asked to answer "agree strongly," "agree," "undecided," "disagree," or "disagree strongly" when given the following statements: "Government should do more to regulate business"; "The best government is one that governs least"; "Government should not interfere in the economy even if unemployment is high"; "Excessive big business profits should be taxed away." A summated index was constructed by adding together the responses to these four questions. We have termed this an economic conservatism scale.
11. Attitudes toward environmental protection: Each respondent was asked to answer "agree strongly," "agree," "undecided," "disagree," or "disagree strongly" when given the following statements: "In order to reduce pollution, the government will have to introduce harsh measures since few people will regulate themselves"; "I am willing to make personal sacrifices for the sake of slowing down pollution even though the immediate results may not seem significant"; "Even if public transportation was more efficient than it is, I would prefer to drive my car to work"; "Pollution is not personally affecting my life"; "The government should provide each citi-

zen with a list of agencies and organizations to which citizens could report grievances concerning pollution"; "The currently active antipollution organizations are really more interested in disrupting society than they are in fighting pollution." A summated index was constructed from the responses to these six questions. We have termed this an environmental support scale.

9 Alternative Hypotheses for Local Energy Innovation

Gerry Riposa

Many observers and researchers interested in the energy crisis have pointed out the potential of local governments to help ease the tensions between growing energy demands and resource constraints (Harrison and Shapiro 1979; Cigler 1981; Sharp and Brunner 1980; Lovins 1977). Support for this position is based on well-documented examples of successful reductions of energy-conservation programs in locales such as Davis, California; Seattle, Washington; and Portland, Oregon. Noticeably absent from explanations of local energy innovation is any analysis regarding the sporadic patterns of local innovation responses (Cigler 1981). This leaves the policy area with various unanswered questions: "Why are some communities innovating in the energy field while others are not?" Or, "If a community decides to implement energy innovation, what determines the final selection of alternatives?" One step toward a framework that produces explanations for the adoption of local energy innovation involves developing testable hypotheses that may yield added information on this policy process. Yet little research has been done in this area to derive them. One promising way to provide insights regarding energy-policy innovation is to examine the literature on policy innovation more generally. Based on a survey of this literature, six hypotheses are suggested in this chapter that should provide researchable departing points as to how and why particular local energy innovations proliferate and are adopted by local decision makers. Prior to discussing these hypotheses, it is necessary to begin by defining innovation and illustrating how an analysis of various innovations may be employed to suggest significant variables in the local energy innovation process.

Innovation: The Definitional Terrain

The term *innovation* has a myriad of definitions. Innovation sometimes is conceived as a new idea, stressing the aspect of inventiveness (Rogers 1962, p. 76). Any number of energy innovations such as solar devices and geothermal development fit this type of conceptualization. This definition has been expanded, incorporating the notion that innovation is the utilization of a process or product new to the organization and introduced from within (Rowe and Boise 1974, p. 284). Los Angeles's decision to implement peak-load pricing for commercial establishments illustrates this strain of innovation definitions.

Nevertheless, these definitions are overly restrictive. For instance, they necessarily exclude innovation brought about by exogenous sources or the adoption of production or processes that have been in the field for a long time but are still new to the adopting institution.

Relaxing the criteria of inventiveness and internal sponsorship, Mohr (1969, p. 112) holds that innovation is the successful introduction of means and ends that are new to the situation. Along the same line, Zaltman, Duncan, and Holbeck (1973, p. 9) define innovation as merely a new process to the existing form. More flexibility is gained from these definitions. Using this less-restrictive definition, cities that adopt federally encouraged thermostat control for public buildings or older conservation measures from adjacent communities fall under local energy-innovation policy research.

From a more-radical perspective, Marxists contend that an innovation is some technical change (Mollenkopf, 1978, p. 130). That is to say, within the capitalist system, a product, process, or idea is an innovation if it increases the political power of the capitalist class and, in doing so, contributes to the perpetuation of capitalist relations in society. Newness and sponsorship have little importance here. It is the political implications that define whether something is an innovation. For instance, when local government establishes a citizen advisory committee for local energy policy, Marxists would contend this perpetuates the system of domination. Like the community-action agencies during the War on Poverty years, these advisory committees are seen as instruments that give the public only the feeling of participation rather than having any real decision-making power. Having no real power, these committees are only a symbolic gesture to give the populace a sense of participation in public affairs in order to displace their anger. Because a citizen advisory committee on energy policy is seen as deflecting public activism, its existence contributes to the retention of power in the hands of private interests and local government elites regarding energy-policy decisions. Consequently, citizen advisory committees are innovations.

The rightness or wrongness of these definitions will not be debated here. Their purpose is to illustrate the denotative range of the term *innovation*. For this chapter, innovation is defined as a process, product, or idea new to the organization, whether introduced internally or externally.

Some Approaches to Innovation Research

Innovation research organizes itself in three approaches that are useful for the examination of the local energy-innovation process. One approach taken by innovation research suggests that the rate of adoption and diffusion is linked to the peculiar characteristics of the individual innovations. Among these characteristics are complexity, cost, and compatibility (Rogers 1962, p. 121). Fliegel and Kivlin's (1966) study of the innovation characteristics that influenced the adoption of modern farm innovations serves as a classic example of this type of research. They argue that the attributes of farm innovations, such as initial cost and complexity, are the critical factors in shaping the farmer's decision to adopt. In relation to local energy innovation, this approach suggests that local adopters may be less inclined toward innovative programs in conservation because of the complexity involved in implementation. A supply-oriented program where the adopter simply locates energy suppliers and buys enough resources to meet consumer needs might be costly but is simple in conception.

Rather than emphasizing the attributes of the individual innovation, a second approach of innovation research stresses the influence of the adopters and their organization. Here the decision maker and the organization are given center stage; they are the independent variables that determine the rate of adoption and the diffusion of innovations (Rowe and Boise 1974; Bingham 1976; Bingham et al. 1981). The adoption decisions are seen as rooted in the decision makers' training, values, and perceptions (Warner 1974). Bingham, for example, calls attention to the importance of the adopters' organizations (1976) and professional associations (1981) in the innovation adoption process. In either case, this genre of innovation research stresses the quality of compatibility between the adopter and the new idea or product. For example, if adopters believe local governments should have a strong welfare function, they would be more receptive to a program such as Boston's loans to low-income families to retrofit their homes for increased energy conservation.

A third approach of innovation research examines adoption of innovations as it pertains to the larger struggle for power. Rather than focusing on the microcharacteristics of innovations or their adopters that singly or together act as the catalyst for adoption decisions, this conception emphasizes the innovation's potential for aggrandizing the adopter's power as the key to its acceptance. Supporting this premise, Galbraith (1973, p. 148) argues that the technostructure adopts innovation to promote its own sovereignty and power. Similarly, Marxists view decisions to adopt innovations in private industry as an attempt by owners of capital to assert their control over the labor force. Indeed, markets and increased profits are also incentives to innovate; however, without first gaining the power to organize and control the labor force, owners would find, according to Marxists, that their profit margins were threatened by an uncoordinated work site and continuous worker demands. By adopting innovations in mass-production technology, factory owners can more readily integrate unskilled labor into the organization of the work force, thus increasing their labor pool. Of tantamount importance, this ability to replace easily workers who need little training or to remove workers altogether gives owners greater power in organizing the work site and confining worker militancy.

To apply this argument to the adoption of local energy innovations, it is suggested that local governments adopt innovation to sustain, if not increase, their power over their constituencies. By adopting energy programs, local governments limit their dependence on other levels of government, while legitimating their authority with their constituents. A successful innovation, such as reducing the use of a city's automobile fleet, can increase revenues for other programs affecting other interest groups, as well as demonstrating to the general constituency that the local administration is doing something to meet energy needs. Both of these tangible effects increase public support for the local administrator. Hence, the political power to govern is maintained.

Alternative Hypotheses

Hypothesis 1: If an energy innovation requires only a variation in the present local political structure, it is more likely to be accepted.

A number of innovation studies suggest this hypothesis. Norman (1971, p. 205), in his study of corporate innovation, indicates that companies tend to adopt innovations in products if they require only a variation, rather than a reorientation, in the existing organizational framework. According to this research, the key factor in the adoption of new products or processes in canned food, medical goods, and autohydraulics was whether the products under consideration required a major change in the existing organizational framework or production process. If they did not, they were likely to be successful innovations in that they were accepted. If, on the other hand, the prospective innovations indicated change in the existing structure, then resistance would be manifested.

Variation, as it pertains to adoption of innovation, means that the adopting entity is required to make little, moderate, or no change in organization to incorporate and utilize the particular innovation. Normally the innovations associated with moderate variations are easy to implement. Low-level programs, such as President Ford's "Whip Inflation Now" buttons to fight inflation, or local governments' fight against energy consumption by adjusting thermostats in public buildings, are examples of these types of policies. While these types of innovations may or may not be effective, the adoption process typically entails new risks or costs or organization change in the existing structure.

Reorientation means that the innovation, if adopted, mandates major, perhaps drastic, change in the adopting organization. If a local government chose to enact an energy ordinance to reduce private consumption of energy during peak-load hours or to require energy-efficient building codes, major policy and organizational change would occur. Planning departments would need to increase measuring procedures for energy flows. Permit processes would become more complex. Staffing requirements would rise if more inspectors were needed. And bureaucracies would gain more discretion in penalty and waiver decisions, thus increasing their authority in local energy policy. Suppose a local government adopted an energy innovation such as a citizen committee to advise them on energy policy. If this were the case, then decision-making processes might slow down due to increased public access. Along with a slowing of policy decisions, one may also expect increased conflict and local power struggles. Both of these occurrences may shift the balance of power in energy policymaking. Because some innovations create this type of reorientation to the existing policy domain, they inspire less receptivity than those that require simple variations.

Studies of local energy innovation support Norman's argument regarding the difference between variation and reorientation in local energy policy adoption. In Cigler's (1981, p. 473) study of adoption of energy programs in North Carolina municipalities, she notes that when these communities did overcome their reluctance to innovate, they usually chose an internal variation such as adjusting thermostats in public buildings. A 1980 survey of California cities and counties found the same avoidance of innovative measures that might dispose local governments toward change. The report identifies only ten local governments in the state that had implemented energy ordinances (CEC, 1980, p. 9). In a 1981 survey of California cities and counties, it was shown that although respondents usually had adopted some energy innovations, they were predominantly measures such as interior lighting adjustment or conservation in the vehicle fleet rather than energy ordinances or production of renewable resources (Office of Appropriate Technology 1981, p. 8). Ostensibly then, when local governments do adopt an energy innovation, they appear to be amenable to ideas that constitute only a variation in their organization or structure.

Of course, it is possible to find showplace communities where a wide range of energy innovations have been implemented, such as Davis, Seattle, and Minneapolis (Brunner 1980, p. 73). These municipalities have experimented with innovations that necessitated both variation and reorientation. Hence, they serve as examples of the potential for a wider latitude of energy-innovation adoptions in the future. For now though, the prospects for adopting a particular energy innovation appear to be substantially influenced by the propensity to provoke organizational and political change.

There is a relationship between the variation-orientation notion and nonincremental policy in the energy-innovation process. Policymaking of a nonincremental nature fails to fit neatly in either the variation or reorientation categories. As Schulman (1975, p. 1355) notes, nonincremental policies are characterized by their indivisibleness. He means that nonincremental policy is a large, often drastic, policy that affects the whole community, where the utilization of the policy's benefits by one citizen does not reduce the ability of another citizen also to enjoy the same level of benefits. Several examples of this type of policy are Davis's vigorous community conservation-education program for residential owners and Portland's offer to inspect residential and commercial buildings for energy efficiency. Although these policies are extensive community programs that demand increases in staffing and expenditures, they present no real change in the political structure. Because this type of program may exceed the incrementalism of variation and yet fails to require the organizational changes of reorientation, the explanations for nonincremental local energy policy may fall outside of the explanatory power of the variation-reorientation hypothesis.

Nevertheless, the variation-reorientation hypothesis draws attention to the possible underlying political forces that may determine why certain innovations are chosen over others. The plausible positive relationship between innovation adoption and low-level structural change may indeed explain why only 29 percent of California's cities and counties have established energy advisory committees (Office of Appropriate Technology 1981, p. 11).

Hypothesis 2: If local government is organized according to the traditional bureaucratic model, it is more likely to adopt energy innovations.

This hypothesis runs counter to the pervasive stereotype of indifferent, lethargic bureaucratic structures that ignore, neglect, and at times resist innovation for fear of assuming a new responsibility. Instead it is the very presence of the bureaucratic structure with its standardization of rules, hierarchy, and division of labor that are seen to increase the probability of innovation adoption. Because the bureaucracy strives to rationalize combinations of knowledge, personnel, and facilities, its decision-making system has a higher range of learning capacity to survey and understand critical factors in its environment (Deutsch 1966, p. 165). Having a higher learning capacity, a bureaucratic system is able to sense, read, and interpret signals of negative feedback and other tension-generating inputs, thus allowing the bureaucracy a sounder basis for responsive policy adaptations (Neiman 1977, p. 9). Further, Neiman argues (p. 9) that this higher learning capacity increases the system's potential for innovative responses since it has a greater capacity to assimilate and apply new information. Therefore, it may be that rather than the bureaucratic structure's being an impediment to innovation, it is a conductor.

Historical support for this position is found in our accelerated rate of capital formation and standard of living over the last eighty years (Blau and Meyer 1971, p. 7) Without the simultaneous growth of a rationalized administrative machinery, material technology and mass production methods would have been inoperable. Bureaucracies routinized and rationalized the innovation that allowed factories with thousands of workers to operate smoothly at a profit while integrating the latest technical methods. If the bureaucratic structures had resisted innovation, then the vast improvements in communication, transportation, and commerce would have been stunted. Not only have bureaucracies contributed to the adoption of innovation, many bureaucracies, like the Nuclear Regulatory Commission, have promoted innovations in the area they are authorized to regulate. This disposed them toward innovation beneficial to the industry. Consequently it is arguable that the bureaucratic structure, far from being an implacable obstacle, is often a prerequisite to the full exploitation of an innovation (Blau and Meyer 1971, pp. 6, 7).

Research on organizational innovativeness supports the positive relationship between the bureaucratic structure and innovation in the public sector. As the public bureaucratic structure promotes rational order and decision making, it cultivates professionalism (Rowe and Boise 1974, p. 268). While professional bureaucrats differ in their functional assignments, they find commonality in similar standards, goals, rules, and procedures. Through this rationalization of the professional bureaucrat, the bureaucratic structure advances the dissemination of new ideas. Operating from broad, yet similar, professional values and goals, bureaucrats transmit innovative techniques among various agencies through trade journals, conferences, and informal dialogues.

Dye (1975, p. 323), agrees, finding that professionalism in bureaucrats and legislators is the most significant stimulus in innovative state policy. Constantly encountering new ideas, the bureaucrat is motivated to experiment with new ideas in order to earn peer-group approval. The more bureaucrats become resources of knowledge, the greater their latitude for implicitly setting standards of acceptability. According to Dye (p. 324), professional bureaucrats are inclined to adopt innovation for reasons of personal power and a sense of self-importance in policy leadership.

While support for the relationship between bureaucratic structure and innovation appears plausible, almost no research has been done on the extent of the bureaucratic structure's influence on local energy innovation. One recent contribution to clarifying this relationship is found in Cigler's (1981, p. 474) North Carolina study. In this research, she discovered that various types of municipal officers—city manager, utility director, safety inspector—had little affect on the level of local energy innovation activities. Meanwhile Bingham et al. (1981, p. 15) have argued that professional associations play a significant role in municipal policy innovation. Because local professionals gain their status and peer-group approval from their competence in the field, they have a long-standing personal interest in increasing productivity for their city. To find new ways of being productive, professional groups like the International City Management Association and the American Public Works Association have historically aided municipalities in innovation (Bingham et al. 1981, p. 16). Under the bureaucratic-structure hypothesis, this professional norm of productivity through innovation would be explained as a direct result of the bureaucratic structural framework.

Even if the bureaucratic structure is not the primary cause of local energy innovation, it may provide useful information about the media and access points of energy-innovation transmission among communities. Further, it allows researchers to distinguish bureaucratic adoption of innovation from political adoption of innovation. Bingham (1976, p. 217) argues that both of these adoption processes have overlapping, yet separable independent variables. Blurring two adoption processes clouds the analysis.

This hypothesis has some limitations. It places the independent variable, the bureaucratic structure, in a vacuum, separate from the social and political environment. This excludes the bargaining and compromise politics of our pluralistic political system. Further, this hypothesis assumes that the bureaucratic structure promotes rationalization, which positively influences the adoption of innovations; however, the bureaucratic structure at times may exhibit irrationality from bureaucratic intransigence or conflict over overspecialization. The point here is that the influence of the bureaucratic structure may have varying cycles that need to be identified and characterized. Finally, one is left with the question: Why are particular innovations selected over others in a network environment of professionals where presumably everyone has similar values? Of course, other limitations are also possible. Nevertheless, it still remains important to measure the extent of influence the bureaucratic structure exerts on local energy policy innovation.

Hypothesis 3: Communities are more likely to adopt energy innovations compatible with the community's needs, values, and resources.

One might argue that this hypothesis gives the local adoption process an issue-specific nature. Rather than correlating the rate of adoption of local energy innovations to some universal characteristic, this proposition suggests that the rate of adoption is a function of the individual community's needs, values, and resources. How a community perceives the impact of innovation on its needs, values, and resources will, of course, affect the probability of adopting it.

Although not dealing with innovation at the local level, Gray (1973, p. 1185) found that an issue-specific factor was influential in the adoption of policy innovation at the state level. Those states that adopted innovations in education were often late or nonadoptive in civil rights and welfare. Menzel and Feller (1977, p. 534) discovered a similar situation in their study of adoption and diffusion of innovative fluoride policies among states. They observed that early adopters of new fluoride programs were not necessarily predisposed to innovation in other areas.

Generalizations about local energy innovations from studies of state policy are vulnerable to a form of ecological fallacy. Yet this research does suggest the possibility that local decision makers, like their state counterparts, assess new activities according to some individual matrix of the innovation's compatibility with the particular community's values, needs, and resources.

There is a certain logic in this hypothesis. When Seattle was faced with a huge investment in a nuclear plant to meet increased energy demands, it chose to implement new local energy-conservation programs that produced increases in available energy resources but were more compatible with their budget. In this case, local fiscal constraints determined Seattle's choice and adoption of energy innovation. In other local situations, different variables may be the overriding influence in a local adoption decision.

Geography, for example, shapes adoption decisions and choice of innovation modes in certain municipalities. When assessing energy innovations, local adopters must take into consideration such factors as the local climate, proximity to energy resources, and population density. Since northern California cities have greater accessibility to geothermal and hydroresources, it is not surprising that Palo Alto, Lompoc, and Santa Clara are inclined toward this type of renewable-resource development (CEC 1980, p. 16). In southern California, warmer climates induce cities like Oceanside, San Diego, and Del Mar to adopt solar ordinances for swimming pools and, depending on location, solar-access mandates for new residential construction (CEC 1980, p. 9). High-density population makes it possible for Los Angeles to establish a methane plant, which demands large volumes of landfill and sewage to be cost efficient. These few examples demonstrate the impact of local factors on the innovation adopter process.

Other localized characteristics shape communities' decisions to adopt an innovation. Suppose a community's cultural perception sees economic prosperity tied to increases in technology and energy production. Any effort to establish a new conservation program may symbolize a retreat from economic growth. Minority constituencies are especially sensitive to any sign of economic slowdown, believing it portends contraction of opportunities. These perceptions damage the chances of adopting local energy-conservation innovations or new, small-scale energy technologies in renewable resources.

Agencies like California's Office of Appropriate Technology have argued that conservation in energy can actually mean industry growth (OAT 1981, p. 7). For example, Lockheed cut energy costs 59 percent through conservation measures (CEC 1981, p. 24). This money theoretically is available for reinvestment, which means jobs.

State leadership also affects community receptivity for energy innovation. The California Energy Commission (CEC) has made a strong commitment to direct energy development in the state toward conservation and renewable resources (CEC 1981, p. 7). Under its auspices, California municipalities find the direction, moral support, and technical assistance to experiment with various local energy innovations. Cigler agrees with this point, noting that the lack of state effort to inform and encourage local communities regarding energy alternatives might account in part for the lack of municipal energy programs in North Carolina. One cannot take this argument too far, however. The CEC cites the reluctance of California locales to make use of outside funds for local energy audits of public buildings, even when the funds are available (CEC 1980, p. 2). Still, state government efforts and policies regarding energy innovation are likely to influence the motivation and threshold of local adoption decisions about energy innovations within their boundaries.

No systematic study has tested the proposition that the adoption of energy innovation is positively correlated with the community's compatibility with the particular innovation. Yet some inductive observations do lend weight to this relationship. For example, among locales that have adopted energy innovation, the specific activities are wide ranging. Although most cities in North Carolina simply adjust thermostats, the cities of Oceanside and Sacramento, California, have implemented solar ordinances (OAT 1981, p. 23). Davis, California, has stressed conservation in public use and energy-efficient building codes, whereas Los Angeles has primarily implemented measures to reduce public demand and local government use (CEC 1981, p. 160). This diversity in actual innovation programs suggests that that policies are individualized to accommodate the existing structure of the community's needs, resources, and values. Given this pattern, it seems plausible that the initial choice concerning the adoption of a particular innovation will be made by juxtaposing the prospective program against the attributes of the community's social system.

Although this community-compatibility, issue-specific hypothesis preserves the importance of the individuality of communities in analysis of energy innovation and may even explain the leap-frogging diffusion pattern of local energy innovation currently visible, it suggests prematurely that no general theory exists in local energy innovation. Perhaps this is true. However, other local-level innovations in such areas as police work and accounting have experienced general acceptance. This begs the question, "Is it that no general theory on local energy innovation exists?" or "Is it that social scientists have yet to build one?" Clearly research efforts are too immature to hold out a conclusion.

Hypothesis 4: The greater the compatibility between the local adopter and the prospective energy innovation, the stronger the chance of adoption.

Under the adopter-innovation hypothesis, the rate of diffusion of an innovation and rate of adoption is a function of the adopter's personal compatibility with the new idea. This compatibility is a product of his or her individual social-psychological-cultural perceptions (Thio 1971, p. 59). These perceptions determine the adopter's view of the prospective innovation. Rogers (1962:121) states that the more the innovation is compatible with the adopter's perceptions, the greater the likelihood of adoption.

From this hypothetical relationship, local energy innovation research must assume a more-sociological direction for reliable explanations about the adoption and diffusion process. Previously some researchers relied on the economic theory of innovation diffusion. Using an S-curve analysis, economists theorized that the rate of adoption was a function of profit; the more profitable the innovation and the smaller the investment, the greater the rate of adoption and diffusion. Unlike this economic determinant theory, the adoption-innovation hypothesis places greater emphasis on the adopter's training, interests, and perceptions (Warner 1974, p. 438). These factors shape the adopter's perceptions of the worth or value of a new energy process or product. Thio's diffusion research lends weight to the incapability of the adopter-innovation hypothesis to explain the variance in innovation adoption. For instance, in their study on modern farm innovation, Fliegel and Kivlin (1966) were unable to explain the positive relationship between the rate of adoption and initial cost. Adopter-innovation compatibility based on the adopter's social perceptions provides the needed explanatory power. Because farm innovations were an accepted means of staying competitive in the farming business and in some ways acted as a status symbol, commercial farmers did not view initial cost as an impediment to adopting new farm techniques (Thio 1971, p. 58). Conversely, subsistence or peasant farmers, operating from a different set of social perceptions, viewed initial cost as an insurmountable obstacle to the adoption of modern farm innovations.

Just as social perceptions influence the decisions of farmers, they underlie the decisions of public officials to adopt local-level innovations. In a study of ninety-four local health departments from 1960 to 1964, Mohr (1969, p. 115) demonstrated the relationship between the local decision maker's attitudes and his or her inclination to see new programs in a favorable way. Those bureaucrats who perceived themselves as highly interactive with their peers and more flexible in the scope of public-health programs were more receptive to local innovation. According to Mohr, this research offers support for the linkage between the compatibility of the adopter's social perceptions and the adoption of local-level innovations.

In terms of local energy innovation, Cigler (1981, p. 476) found that the absence of aggressive innovation activity in North Carolina might be related to the majority of respondents (77 percent) who felt that outside interest and assistance from other levels of government were integral to the development of local energy policies. Whether state or federal stimuli are actually needed in local energy innovation is not the point. By perceiving the prerequisites of local programs as outside factors, local decision makers' voluntarily limited the possibilities of adopting energy innovations at their command. To act otherwise would be incompatible with their perceptions.

One further hypothetical example might add clarification. Some cities own and operate their power utilities. California has twenty-two municipally owned utilities headed by a utility director and responsible to the city council. Suppose, in this type of situation, a city council and the utility director believed their responsibility was to produce energy for their constituents while bolstering the general fund. Operating from this type of perception, the city might choose to invest in production technology that has the potential to produce energy beyond their local needs, making it possible to sell the surplus to other producers to increase city revenues. A different city, one with an ecology-minded city council and utility director, might perceive that the best way to increase energy resources and revenues is through adopting innovation that affects the demand side of energy consumption, such as retrofitting homes and creating extensive bicycle lanes.

One obvious problem with this hypothesis is that, like purely economic explanations, it lacks full social dimensions. Warner (1974, p. 430) argues that any comprehensive theory of adoption and diffusion of innovation must view this complex process as a combination of both economic and social determinants. He does not discount the importance of individual traits and perceptions in adoption process; however, he questions sociological exclusiveness in the causal relationship explaining why processes or products are adopted. For stronger explanations of this phenomenon, researchers must analyze the sociological research on the innovativeness of the individual adopter and the economic research focusing on the aggregate of individual adopter decisions. Warner suggests that the key to understanding the adopter of innovations may be found in future research in the interactive effects of these sociological and economic variables.

Hypothesis 5: Community environment variables are the determining influence in the adoption and diffusion of innovation.

Social scientists have established a significant correlation between community environment variables—size, ethnicity, socioeconomic status, suburbia—and public policy (Hawkins 1971; Dye and Gray 1980; Anderson 1981). Research, using policy innovation as a concept, also demonstrates a correlation between community-environment variables and innovation adoption. In a study on local innovation in public housing and urban renewal, Aiken and Alford (1970, p. 863) argued that size indicated a historical assimilation of past innovations, making it more amenable to future innovations. Researchers investigating community innovation in public housing, fluoride programs, and automated services have acknowledged the important interactive effect of size and community need on local innovation (Agnew, Brown, and Herr 1978, pp. 23, 25). Even Dye, who stressed earlier the significance of professionalism in innovation, agrees that an area's environment has a powerful effect on the decision-making process; however, Dye argues that rather than size, it is a community's wealth that indicates a propensity to experiment with innovation. These research findings suggest that a similar causal relationship may exist between community environment and the adoption of local energy innovation.

Although not dealing with energy innovations, Bingham (1976) adds verification to the community-environment variable in his study of local innovation in public housing, special districts, public libraries, and other common government functions. Bingham demonstrates that within the community environment, the community's socioeconomic status accounts for the most variance. The next-strongest effect is suburban development, followed by the level of racial or ethnic segregation. Community size did have some effect on innovation but registered the least impact. Bingham's study showed that size had less direct impact on innovation than it did in creating demands on the community's organizational characteristics and environment, which directly affects the adoption of the innovation. In this sense, the community environment is more the origin of local innovation. This hypothesis supplements Walker's theory that diffusion of innovations in the community spreads by competition and emulation (1969, p. 880). Under the community-environment hypothesis, the stimulus to innovate is generated by community characteristics and other environmental variables, including other local governments.

Local energy-innovation research has yet to test rigorously the influence of the community's environment on the adoption and diffusion of energy programs. Hardly anyone will argue that it does not have some impact, but whether these independent variables assume a predominant role is still quite tentative. For example, communities as small as Fairfield, California, to as large as Los Angeles have established renewable-resource projects (OAT 1981, p. 1). And communities of varying wealth, such as Boston, Seattle, and Oakland, have instituted conservation programs (DOE 1979, p. 11). Research on the adoption process must explain these variations.

Hypothesis 6: The adoption and diffusion of local energy innovation is a local response to the growing contradiction between social production and private ownership in our capitalist system.

According to this hypothesis, local decision makers adopt energy innovations to avoid further federal incursions into public affairs, without jeopardizing the stability of the social system. The federal government, especially since the New Deal, has steadily increased its presence in producing the basic material services consumed by the public at large in a system purportedly driven by market forces and private ownership. According to some analysts, this poses a social contradiction in the logic of our system (Castells 1978, pp. 5, 18). Tensions emerge as private interests and other levels of government seek to adjust the equilibrium of political power in the face of a growing federal presence. It is this contradiction that instigates local governments to innovate to maintain their power and legitimacy with their constituency.

As citizens directly produce less and less of their perceived necessities, they turn to the government apparatus to provide those public goods society consumes collectively, such as public education, transportation, medical services, disaster relief, and housing (Castells 1978, p. 3). Since local governments are unable and private businesses unwilling, the national government has assumed a greater role in the production of these services. Through the use of federal revenues, the national government is able directly or indirectly to produce programs for home mortgages, flood damage, and national innoculation programs against disease, to cite only a few examples. Because the cost of these publicly provided benefits is socialized among the nation's taxpayers, this policy activity is called social production.

Yet it is a mistake to see this growth of social production as some calculated move by a power-hungry Washington. On the contrary, it is a systemic response to promote economic stability by providing the material service necessary for a growing, productive, yet passive, labor force. Due to the connection between satisfaction of public explanations and national social and economic stability, national social production becomes necessary for the system's persistence (O'Conner 1973, p. 124). Yet as in any other public policy, this action has externalities. Increases in social production forge inroads for national power. Once energy became a national-security issue, it proved no exception to this political pattern. As in other areas of social production, the federal government has expanded its leadership role in energy.

The Reagan administration demonstrates both of these qualities in energy production. President Reagan pledges support for one hundred one-thousand megawatt nuclear plants (Daneke 1980, p. 21). James Watt, secretary of the interior, proposes the opening of the national coastlines for oil explorations. The president proposes financing high-technology energy research (Cigler 1982, p. 765). At the same time, decentralized programs with considerable levels of local participation, such as the Residential Conservation Service and the Appropriate Energy Extension Service, have been marked for fiscal extinction (Cigler 1982, p. 766). These energy initiatives show the executive branch's attempts to assert a leading role in energy-production policies. While this type of social production, with its supply-side orientation, may or may not answer the demands for growing levels of energy consumption, it does increase the federal role in local affairs.

Federally sponsored energy policies can create local policy problems that officials have little chance of solving with their own resources. Opening leases for off-shore oil drilling can devastate local coastal communities whose economies are based on fishing and tourism. Greater development of coal under the Power and Industrial Fuel Act of 1978 can increase ambient air qualities for surrounding locales and strain the local resources of the communities experiencing the boom in population from the influx of migrants seeking employment in mining and construction. Faced with these externalities from federal energy initiatives, local governments will have to enact policies in education, pollution, and housing to offset the effects on their communities. Ironically, the local governments will seek at least part of their funding to alleviate the problems from Washington, thereby further increasing federal monitoring and intervention in local affairs.

Presented with this increasing hegemony in local affairs, local governments' attempt to adopt innovations that preserve their sense of autonomy and authority over their constituents. Harrison and Shapiro (1979, p. 27) argue that local governments have successfully demonstrated their ability to implement conservation and renewable resources that can maintain their autonomy. Conservation programs in Boston reduced city expenditures, prompting lower tax levels and providing funds for other areas of collective consumption. Programs of renewable resources in Oceanside and Sacramento may offer businesses clean, cheap energy. Both of these policies fight the erosion of tax bases and may encourage a shift in residential and industrial patterns and preferences. Improving their fiscal status also permits localities to reduce their dependence on federal money, thus minimizing access points of federal intervention. Therefore local governments that innovate to provide the basic necessity for modern living, energy, reduce expenditures and increase revenues, allowing them to maintain their integrity in the federalist system and legitimating their position of power with their constituency.

The major limitation of this hypothesis lies in its capacity to be measured by defining the relationship between social production and local energy innovation. How can one control for the impact of social production on local energy innovation in a capitalist system? Without a rigorous effort to conduct research on this position, it stands an excellent chance of being relegated to the status of a melodramatic conspiracy theory.

On the other hand, the social-production hypothesis does suggest some novel aspects in understanding the local energy-innovation process. First it suggests that local innovation is no longer the predilection of discrete communities but rather local responses to the structural relations in society brought about by the contending forces between social production and private ownership. A corollary value of this premise is that political structure, politics, and power are reintroduced as integral variables in the analysis of energy adoption and diffusion. Second, this hypothesis offers an explanation for the apparent slow start in the adoption and diffusion of innovations. It holds that innovation diffusion is a social process, dependent on the level of social production currently taking place. Energy innovation is not conceptualized along a continuum of linear progression; it is a phenomenon subject to the rise and fall in the tensions inherent in social production. Finally, this hypothesis demands that the researcher examine the energy-policy-innovation process as part of the social phenomena of society.

Conclusion

The purpose of this chapter is to contribute to the development of a research agenda that identifies and explains the causal relationships in the adoption process of local energy innovation. Because little systematic research has been done in this policy area, a necessary step in this theory-building stage is to promulgate researchable hypotheses that purportedly address the relationship between different independent variables and the rate of adoption and diffusion in local energy innovation. To that end, an admittedly partial list of six hypotheses, drawn from the literature on innovation in other policy areas and energy innovation, was advanced. These seem to offer reasonable starting points for the systematic research of local energy innovations. Included in these propositions on the adoption process were generalizations about the independent effects of the individual innovation; the individual adopter; the adopting organization; the community's environment; the community's value structure; the bureaucratic model; and the struggle for power. I hope that future empirical work will determine the extent to which each of these hypotheses, as well as others not discussed here, contributes to explanations of why public organizations like local governments decide to manage energy problems and why there is such a variety in the patterns of adoption.

Of course, one may argue that the development of adoption hypotheses and attempts at theory building on local energy are premature. In other words, such attempts will outpace the existing body of raw data on this policy question. This is true, at least in the second part. Few data on local energy innovation do exist. Nonetheless, this should not be an obstacle into the early activities in theory building. Neiman (1977, p. 24), in his research on analogues for policy analysis, argued that theories prior to raw data signal gaps in our knowledge in a particular policy area, thereby showing future researchers the relevant theoretical needs. The hypotheses set out here should direct research in the study of the adoption of local energy innovations, or at the very least, signal the gaps in our knowledge about the factors that constrain or facilitate the adoption and diffusion of local energy innovations.

References

Agnew, J.A.; Brown, L.A.; and Herr, P.J. 1978. "The Community Innovation Process: A Conceptualization and Empirical Analysis." *Urban Affairs Quarterly* 14 (September): 3–30.

Aiken, M., and Alford, R.R. 1970. "Community Structure and Innovation: The Case of Public Housing." *American Political Science Review* 64 (September): 843–864.

Anderson, D.D. 1981. *Regulatory Politics and Electric Utilities.* Boston: Auburn Press.

Bingham, R.D. 1976. *The Adoption of Innovation by Local Government.* Lexington, Mass.: Lexington Books, D.C. Heath and Company.

Bingham, R.D.; Hawkins, B.; Frencheis, J.P.; and Le Blanc, M.P. 1981. *Professional Associations and Municipal Innovations.* Madison: University of Wisconsin Press.

Blau, P., and Meyer, M.W. 1971. *Bureaucracy in Modern Society.* 2d ed. New York: Random House.

Brunner, Ronald D. 1980. "Decentralized Energy Policies." *Public Policy* 28 (Winter): 71–91.

California Energy Commission. 1980. *Local Government Energy Programs: States, Opportunities, and Recommendations for State Support.* Sacramento: California Energy Commission.

California Energy Commission. 1981. *Energy Tomorrow, Challenges and Opportunities for California: 1981 Biennial Report.* Sacramento: California Energy Commission.

Castells, M. 1978. *City, Class and Power.* New York: St. Martin's Press.

Cigler, B.A. 1981. "Organizing for Local Energy Management: Early Lessons." *Public Administration Review* 41 (July): 470–479.

————. 1982. "Intergovernmental Roles in Local Energy Conservation: A Research Frontier." *Policies Studies Review* 1 (May): 761–776.

Craig, P.P.; Darmstadter, J.; and Rattian, S. 1976. "Social and Institutional Factors in Energy Conservation." *Annual Energy Review:* 535–551.

Crain, R.L. 1966. "The Diffusion of an Innovation among Cities." *Social Forces* 44 (June): 467–476.

Daneke, G.A. 1980. "The Poverty of National Energy Policy and Administration." In *Energy Policy and Public Administration,* edited by G. Daneke and G.K. Lagassa, pp. 9–27. Lexington, Mass.: Lexington Books, D.C. Heath and Company.

Deutsch, K. 1966. *Nationalism and Social Communications: An Inquiry into the Foundations of Nationality.* Cambridge: MIT Press.

Downs, G.W., and Mohr, L.B. 1976. "Conceptual Issues in the Study of Innovation." *Administration Science Quarterly* 21 (December): 700–713.

Dye, T. 1975. *Understanding Public Policy.* 2d ed. Englewood Cliffs, N.J.: Prentice-Hall.

Dye, T.R., and Gray, V., eds. 1980. *The Determinants of Public Policy.* Lexington, Mass.: Lexington Books, D.C. Heath and Company.

Fliegel, F.C., and Kivlin, J.F. 1966. "Attributes of Innovations as Factors in Diffusion." *American Journal of Sociology* 72 (November): 235–248.

Galbraith, J.K. 1973. *Economics and the Public Purpose.* Boston: Houghton Mifflin.

Gray, V. 1973. "Innovation in the States." *American Political Science Review* 67 (December): 1174–1185.

Harrison, D., and Sharpiro, M.D. 1979. *The Local Government Role in Energy Policy.* Boston: Harvard University Press.

Hawkins, B.W. 1971. *Politics and Urban Politics.* Indianapolis: Bobbs-Merrill.

Jones, C.O. 1979. "American Politics and the Organization of Energy Decision Making." *Annual Review of Energy:* 99–121.

Lovins, A.B. 1977. *Soft Energy Paths: Toward a Durable Peace.* New York: Harper and Row.

Menzel, D.C., and Feller, I. 1977. "Leadership and Interaction Pattern in the Diffusion of Innovation among the States." *Western Political Science Quarterly* 30 (December): 528–536.

Mohr, L.B. 1969. "Determinants of Innovation in Organizations." *American Political Science Review* 63 (March): 111–126.

Mollenkopf, J.H. 1978. "The Postwar Politics of Urban Development." In *Marxism and the Metropolis,* edited by W. Tabb and L. Sowers, pp. 117–152. New York: Oxford University Press.

Neiman, Max. 1977. "Analogues and Policy Analysis: An Alternative Strategy." *American Politics Quarterly* (January): 3–26.

Norman, R. 1971. "Organizational Innovations: Product Variations and Reorientations." *Administration Science Quarterly* 16 (June): 203–215.

O'Connor, J. 1973. *The Fiscal Crisis of the State.* New York: St. Martin's Press.

Office of Appropriate Technology. 1981. *Local Energy Initiatives: A Survey of Cities and Counties in California.* Sacramento: Office of Appropriate Technology.

Rogers, E.M. 1962. *Diffusion of Innovation.* New York: Free Press.

Rowe, L.A., and Boise, W.B. 1974. "Organizational Innovation: Current Research and Evolving Concepts." *Public Administration Review* 34 (May): 284–293.

Schulman, P.R. 1975. "Nonincremental Policy Making: Notes toward an Alternative Paradigm." *American Political Science Review* 69 (December): 1354–1370.

Sharp, P., and Brunner, R.D. 1980. "Local Energy Policies." In *Energy Policy and Public Administration,* Edited by M. Daneke and G.K. Lagassa, pp. 83–96. Lexington, Mass.: Lexington Books, D.C. Heath and Company.

Thio, A.O. 1971. "A Reconsideration of the Concept of Adopter-Innovation Compatibility in Diffusion Research." *Sociological Quarterly* 12 (Winter): 56–68.

U.S. Department of Energy. 1974. *Local Government Energy Activities,* Vols. 1–3. Washington, D.C.

Walker, J. 1969. "The Diffusion of Innovation among the American States." *American Political Science Review* 63 (September): 880–900.

Warner, K.E. 1974. "The Need for Some Innovative Concepts of Innovation: An Examination of Research on the Diffusion of Innovation." *Policy Science* 5: 443–457.

Zaltman, G.; Duncan, R.; and Holbeck, J. 1973. *Innovations and Organizations.* New York: John Wiley.

Index

About the Contributors

Eugene Bardach is a professor at the Graduate School of Public Policy, University of California, Berkeley. He has published extensively and has a continuing interest in social regulation, the impact of the media on public policy, and the theory of social cooperation.

Beverly A. Cigler is an assistant professor in the Department of Political Science and Public Administration, North Carolina State University. Her current research concerns local energy-policy development and implementation and state-local policy coordination. She is currently involved in ongoing research about local energy policy.

Thomas Dietz is an assistant professor of sociology at George Washington University. In addition to his research on market and organizational structure in the solar industry, he is participating in a study of residential energy conservation and is preparing a book-length manuscript on theory and method in social-impact assessment.

James P. Hawley is an assistant professor of sociology at the University of California, Davis. His recent work on energy focuses on socioeconomic factors affecting diffusion of solar-energy technology. He is also doing research on the policies of the Federal Reserve Board, monetary politics, and industrial structure.

Karl Hausker holds the M.A. in public policy and is currently a doctoral candidate at the Graduate School of Public Policy at the University of California, Berkeley. He is serving as an intern with the California Public Utilities Commission and is conducting research about utility regulation.

Robert A. Johnston is an associate professor in the Division of Environmental Studies at the University of California, Davis. His research interests concern urban growth and density and their relationship to energy use in transportation.

Walter J. Mead is a professor of economics at the University of California, Santa Barbara. He is widely known for his work on resource economics. Recently he has studied the oil- and gas-leasing systems for the outer continental shelf. He is coeditor of *U.S. Energy Policy: Errors of the Past, Proposals for the Future.*

Gregory G. Pickett is a visiting lecturer doing postgraduate work at the University of California, Santa Barbara. His research interests include the economics of government regulation, natural resources, and financial economics.

Michael D. Reagan is a professor of political science at the University of California, Riverside. His research interests are energy and resources policy and government regulation. He is a leading figure in the study of intergovernmental relations.

Gerry Riposa is a doctoral candidate in political science at the University of California, Riverside. He is currently conducting research on the impact of state energy policy on local government. His major fields of study include public policy, urban politics, and comparative politics.

Michael F. Sheehan is an assistant professor in the Department of Urban and Regional Planning at the University of Iowa. His current research interests include energy policy, public utility regulation, environmental planning, and local economic development.

Steve Tracy is a doctoral student in the Division of Environmental Studies at the University of California, Davis. His main research interests are the siting of nuclear power plants and implementation of general plans by local government.

About the Editors

Max Neiman is an associate professor of political science at the University of California, Riverside. He is the author of numerous articles about urban politics. His current research interests include locally adopted energy policy, land-use regulation, environmental politics, and urban theory.

Barbara J. Burt is a doctoral candidate at the University of California, Riverside. Her recent research has included an examination of local government adoption of energy policy.

About the Authors

Date Due

Demco 38-297